Springer Mineralogy

More information about this series at http://www.springer.com/series/13488

Jiashu Rong · Fenggang Wang

Metasomatic Textures in Granites

Evidence from Petrographic Observation

Jiashu Rong
Beijing Research Institute of Uranium
    Geology
Beijing
China

Fenggang Wang
Beijing Research Institute of Uranium
    Geology
Beijing
China

ISSN 2366-1585   ISSN 2366-1593 (electronic)
Springer Mineralogy
ISBN 978-981-10-9223-7   ISBN 978-981-10-0666-1 (eBook)
DOI 10.1007/978-981-10-0666-1

Jointly published with Science Press Ltd., Beijing, China

Printed on acid-free paper

This Springer imprint is published by Springer Nature
The registered company is Springer Science+Business Media Singapore Pte Ltd.

Foreword 1

The book "Metasomatic Texture in Granites" is an atlas, in which the authors stress on the exposition of metasomatic textures and their formation mechanisms than other atlases both at home and abroad. In the book the metasomatism of various minerals, including albite, K-feldspar, muscovite, biotite, quartz, beryl, calcite, apatite, etc., is systematically discussed with clear, colored, microscopic pictures and explanations.

As a new attempt and development, the authors advance both hetero-oriented and co-oriented patterns of monomineral replacement in granites. The hetero-oriented replacement pattern includes the albitization and the K-feldspathization occurring at grain boundary of plagioclase with K-feldspar and of two K-feldspars, the muscovitization and biotitization at grain boundary of K-feldspar with mica and of two biotites, etc. The co-oriented replacement pattern includes mainly the deanorthitization of plagioclase, co-oriented albitization of K-feldspar, co-oriented muscovitization and chloritization of biotite, etc. The classification of mineral replacement reflecting the objective facts is of important significance.

The authors elaborate systematically the formation mechanism of mineral replacement, its genetic hypotheses, the passage for hydrothermal fluids, the replacement mechanism (dissolution-precipitation and ion-exchange), etc. Myrmekite is the most visible texture in granite. So far there are six genetic hypotheses about its formation mechanism, including nibble replacement. After quoting extensively to discuss the possibilities and contradictions of the other hypotheses, the authors prove that myrmekite is actually formed from K-feldspar replaced by Na plus Ca bearing fluid, forming the replacive plagioclase leaning against a feldspar crystal either plagioclase or K-feldspar. As the SiO_2 content needed to compose replacive plagioclase is less than that contained in the replaced K-feldspar, the surplus SiO_2 remaining in situ produces the vermicular quartz grains.

The book is written by Prof. Jiashu Rong on the basis of ample material collected by all means after his retirement with the encouragement given by Prof. Lorence Collins. Rong hopes that his knowledge and judgement obtained in his study on metasomatic textures in granites reflect the objective situation. However, whether it is a perfect reflection requires practical inspection and extensive review by all experts on the line. There might be different viewpoints about some phenomena for our further discussion and in-depth study before we get a scientific explanation.

The research on metasomatic textures in granites is a supplement for igneous petrography. Moreover, it is of valuable reference to the study of formation mechanism of granite. I hope the book would be published as soon as possible, so as to promote an intensive communication and a deeper study on it.

May 2015   Qihan Shen
                                                       Academician of the Chinese
                                                         Academy of Science

Foreword 2

The monograph "Metasomatic Texture in Granites" is a research achievement of researcher Dr. Jiashu Rong and his team through their hard working for years. The phenomena described in the book can commonly be found by those who deal with identification of granite and gneiss.

Based on detailed observation on granite, the authors emphasize that metasomatic textures are basically classified into two patterns: (1) hetero-orientation (nibble) replacement; (2) co-orientation replacement. This brand-new idea is helpful to understanding of metasomatic texture under microscopic observation.

I had no concept of co-orientation or hetero-orientation in my teaching and research work in previous years, so I could not clearly explain the metasomatic texture in granites. For instance, some plagioclase inclusions in K-feldspar phenocryst have got clear rim while the others have not. I had no idea to explain it without consideration of coherence of the crystallographic orientation between plagioclase with K-feldspar. After reading the book, at once I have become clear minded.

Besides clear rim, many problems are involved in the book, such as myrmekite, perthite, antiperthite, especially the historical analysis of metasomatic processes, which provides a new viewpoint on the superposition of multiple metasomatic processes in granite and on the relationship of rare metal mineralization with metasomatism in granite.

Petrologists have been discussing the origin of myrmekite for 140 years since its discovery and had been formulating various presumptions on it. In this book, many examples are enumerated, indicating that myrmekite appears mainly in the rim of plagioclase contacted with differently oriented K-feldspar as well as at the boundary between two K-feldspars, forming the swapped myrmekite. The authors also discovered the existence of residue of perthitic albite, even K-feldspar with their original orientation preserved in myrmekite. Therefore, it gives a firm support to the theory of hetero-orientation metasomatic genesis of myrmekite.

It is appreciable that the authors submit their academic ideas on metasomatism on the basis of a scientific analysis to the viewpoints of previous geologists, so as to create a good atmosphere for an academic exchange. For example, the authors advance their practical ideas on disputed areas of granite as the genesis of perthite, the difference between phenocryst and porphyroblast, etc. The authors also make an analysis on the genesis of chloritization of biotite in accordance with data referring to high-precision transmission electron microscopic images, and convincingly point out the two kinds of mechanism for chloritization. Though the question of whether plagioclase may be replaced by co-oriented K-feldspar is still in debate, in the book the authors also refer it with experiment data published by Lobatka (2004), thus bringing a new way toward its resolution.

In brief, the authors show us the phenomenological characters of meta-somatic texture in an all-round way by analyzing them cautiously and systematically and making a penetrating explanation, thus guiding us to see the essence through the phenomena. So it is a valuable monograph about petrography, which is of great practical reference to the people dealing with the identification of rocks and minerals and the study on granite-related rocks.

Origin of granite has been a puzzling topic in the field of geology, while the study of metasomatic texture in granite is exactly one of the keys to the mystery. As far as I know that the anatexis has been regarded as a main factor of genesis of migmatite at present. In this regard, the monograph is surely interesting for those who engage in research on either magmatic granite or metamorphic rocks.

June 2015   Zhendong You
                                                    China University of Geosciences

Foreword 3

Dr. Jiashu Rong is a senior petrologist in our Institute.

Since 1970s he has been making creative achievements in petrological study of granites in Southern China (uranium ore-field of Zhuguang, Guidong, Taoshan, Huichang, Jingtan, Motianling, etc.), in application of the method of induced fission tracks in uranium geology, in the survey of uraninite of 300 granite bodies in China, in the study of uraninite deposits of pegmatic granite of Hongshiquan and Danfeng, as well as in the petrological study of mantle xenoliths sponsored by China Natural Science Fund. In recent years he has been devoting his whole mind to metasomatic textures in granites. I think the following major points he proposed deserve to be emphasized.

1. Metasomatic texture in granites should be divided into hetero-oriented and co-oriented replacement patterns. The former is predominant. Its formation mechanism is of dissolution–precipitation. K-feldspar is the most easily replaced mineral for hetero-oriented albitization, K-feldspathization, quartzification, and muscovitization. The formation mechanism of co-oriented replacement of biotite by chlorite or muscovite is of ion-exchange. As for the co-oriented albitization of plagioclase or K-feldspar, the formation mechanism is to be explored.

2. The process of hetero-oriented replacement does not occur at random. It is controlled by two prerequisite conditions: a mineral prone to be replaced on the one side, and a same or similar kind of replacive mineral on the other side as nucleation substrate for replacive growth. Only when metasomatic process should have occurred, while the same or similar kind of the replacive mineral is absent, then an impurity may act as a nucleation substrate for the growth of a replacive mineral.

3. Co-oriented albitization is easy for plagioclase, but hard for K-feldspar. However, once K-feldspar is co-oriented albitized, it happens so rapidly and thoroughly, that the transition zone would be hardly observable.

4. The albite formed by co-oriented albitization of K-feldspar and the perthitic albite may still be distinguished by their different forms and occurrences.

5. K-feldspar may be subjected to both hetero-oriented and co-oriented albitization. Although the former develops intensely, the latter does not happen at all. On the contrary, when the latter develops strongly, the former is still not enhanced. It implies that not only they took place

successively and individually, but also their formation conditions and environments are quite different. Therefore, it is reasonable and necessary to divide them and describe them separately rather than lump them together.

6. Inserting a quartz plate under crossed polars is a helpful procedure to determine the consistence of crystallographic orientations of tiny relics with surrounding minerals, in order to identify either hetero- or co-oriented replacement phenomena.

7. The small platy albite minerals in Li-F granites are of primary rather than "chaotic" metasomatic in origin.

8. Myrmekite is produced from K-feldspar replaced by plagioclase, as the SiO_2 content needed to compose replacive plagioclase is less than that contained in K-feldspar and the surplus SiO_2 remaining in situ produces the vermicular quartz grains.

Besides, he has got the following tendentious conclusions, such as (a) Quartz can not be replaced by alkali-feldspar (albite or K-feldspar); (b) Similar to the condition in magma, the growth ability of replacive albite along the axes a and c is much greater than along the axis b; (c) Perthitic albite in K-feldspar is mainly formed by either unmixing of solid solution or simultaneous crystallization of K-feldspar rather than replacement origin; (d) The idiomorphic interlocking crystal of K-feldspar with albite is primary rather than metasomatic origin; (e) Megacryst of K-feldspar in granitic rocks is basically of phenocryst or its residue rather than porphyroblast; (f) It is possible to distinguish the successive sequence of multiple superimposed metasomatic processes according to the rule of mineral replacement.

What is mentioned above is only my understanding. And I believe in the book there must be many valuable viewpoints for readers to find.

In conclusion, the book represents a creative research achievement on metasomatic textures in granites. It is a rare high-level petrography monograph in recent years and worth to publish in both Chinese and English languages.

July 2015   Letian Du
                                 Beijing Research Institute of Uranium Geology

Preface

After completion of crystallization from a magma, under subolidus at about 500 °C along with slow decrease of temperature and reaction of residual fluids, the primary minerals in granite may undergo a change either in composition or in structure. Thus, K-feldspar may exsolve albite lamella and may be transformed from orthoclase into microcline. These changes are of subsolidus re-organization, but not metasomatic phenomena, and the bulk compositions are not modified.

However, after formation of granite, the hydrothermal fluid, either relevant or irrelevant to magmatic processes, can penetrate into solidified granitic rocks, resulting in the instability or change of an individual primary mineral followed immediately by precipitation or crystallization of a more stable new mineral, thereby partly changing the primary mineral into a new one.

Lindgren (1925) stated: "Replacement in solid rocks consists in solution of the host mineral, followed immediately by deposition of an equal volume of the guest mineral......The volume of the replacing mineral equals the volume of the mineral replaced. Deposition follows so closely upon solution that at no time can any open space be discerned under the microscope."

Mineral replacement, such as cation exchange, deuteric alteration, as well as pseudomorphism, chemical weathering, leaching, diagenesis, and metamorphism are all linked by common features in which one mineral is replaced by a more stable one (Putnis 2002).

Besides, under new physical–chemical circumstances, the compositional transformation of an old mineral locally or wholly into a new one by ion-exchange, without dissolution–reprecipitation, is also a reaction product of metasomatism.

Generally speaking, the metasomatic process is characterized by the following features:

1. A rock remains in solid state as the whole metasomatic process proceeds.
2. Transformation (reformation) or dissolution of a previous mineral occurs almost simultaneously with formation of regenerated minerals without any evidence of open space.

3. The volume of the replacive guest mineral is equal to the volume of the host mineral that is replaced.[1]

The process and the phenomenon of a primary mineral replaced by a newly formed mineral are always indicated by the name of replacive mineral plus -ization or -ification. For example, the replacement of K-feldspar by albite is called albitization of K-feldspar; the replacement of biotite by chlorite is chloritization of biotite, the replacement of K-feldspar by quartz is quartzification of K-feldspar, etc. The degree of replacement, i.e., the degree of -ization or -ification may be described as weak, medium, intense, and complete.

If one mineral is partially dissolved by one solution, resulting in presence of a free space where new minerals may be deposited later rather than simultaneously, this process should not be treated as a replacement but as a type of filling.

Metasomatism is conceptionally different and distinguished from the isomorphism that occurs during magmatic crystallization. Metasomatic phenomena are similar to cotectic ones when two minerals simultaneously crystallize, so they may often be confused. Moreover, metasomatism is easily indistinguishable from unmixing (exsolution) in a solid solution.

Key evidence of replacement is the presence of relics of replaced minerals in the replacive mineral. However, such relics should be distinguished from the crystals formed by simultaneous crystallization and from normal inclusions in igneous rocks. The relics also should be determined as a real residue from the replacement seen at present rather than an inherited relict from the previous replacement.

On the basis of hundreds of thin sections from granites in China, new recognition of metasomatic phenomena has been obtained. According to the consistency of orientations between replaced and replacive minerals, metasomatic textures are basically classified into two patterns: (1) hetero-oriented replacement; (2) co-oriented replacement. Hetero-oriented albitization of K-feldspar is quite distinct from co-oriented albitization of K-feldspar. They occurred separately and individually without a transition. Their formation conditions must be different from each other, although both of them generally are all called albitization of K-feldspar.

Origins of clear albite rim, intergranular albite, small platy albite, K-feldspathization, quartzification, chloritization, muscovitization, beryllization, myrmekite (myrmekitization), perthitic albite, antiperthite, as well as K-feldspar megacryst are comprehensively discussed and reasonably explicated. Metasomatic regulations and superimposed metasomatic processes are clearly illustrated by a series of color microphotos taken with quartz plate inserted under crossed polars.

---

[1]There is another point of view for "unequal volume replacement": during metasomatism, the introduction of a large amount of K_2O, Na_2O, SiO_2 is proceeded without carrying out of the matching volume of CaO, MgO, FeO, Fe_2O_3, resulting in upswelling and transformation of geosyncline sedimentary rocks to large-scale alkaline-rich acidic granite bodies (Geological department of Nankin University 1981; Granitoids of various ages in southern China and their relationship with metallogenesis (in Chinese) p. 381).

The metasomatic phenomena discussed in this book apply to single individual mineral replacements in granites. The recrystallization and the replacement of a primary mineral by an aggregate of new minerals are more complicated and not discussed here.

Beijing Research Institute of Uranium Geology   Jiashu Rong
February 2015   Fenggang Wang

References

Lindgren W (1925) On metasomatism. Bull Geol Soc Am Bull 36:247–262
Putnis A (2002) Mineral replacement reactions: from macroscopic observations to microscopic mechanisms. Mineral Mag 66:689–708

Acknowledgements

The authors are grateful to Beijing Research Institute of Uranium Geology for encouragement and support. Much appreciation goes to Division of Geology and Mineral Resources of the Institute for logistical support. The authors also acknowledge the logistical help for field work from Foshan Geological Bureau, Guangdong Province. The authors are indebted to Academician of Chinese Academy of Sciences Qihan Shen and Dr. Letian Du for kindly expressing their financial support to start the project. To Drs. Shijie Huang, the late Weixun Zhou, the late Yueheng Guo, Jiashan Wu, Honghai Fan, Jianguo He, Jinrong Ling, Jianzhong Li, Tianfeng Wan, Fuyuan Wu, the authors extend heartfelt thanks for their concern, encouragement, and help. The late Dr. Yanting Wang, Drs. Zhifu Sun, Zhizhang Huang, Xiheng Fang, Wenguang Wang, Youxin Xie, Genqing Cai, and Xiuzhen Li of BRIUG are thanked for collaboration, discussion under microscopic observation. Many thanks go to Academician of Chinese Academy of Sciences Chongwen Yu, Professors Zhendong You, Jiaxiang Qiu, Xunruo Zhou, Shanping Sun, Fengxiang Lu, Jiashan Wu, Dawei Hong, and Yongshun Liu for attentive review and precious comments of the manuscript. The authors thank researcher Yuexiang Li, engineers Donghuan Chen, Hao Xu, Yuanqiang Sun, and Qiang Xu for their assistance in computer application. Engineer Yanning Meng is thanked for revision of sketch drawing and partial translation of the manuscript. Lingling Yuan, the senior author's wife, is appreciated for her loyal encouragement and improvement of English.

Professor Lorence Collins, Department of Geological Sciences, California State University-Northridge, has been studying metasomatic phenomena in granites and shows his unique ingenuity in discussing the origin of myrmekite. The senior author had not known him until the senior author's short essay about metasomatic texture was transmitted to him in 1987 by an American participant in Guangzhou International Granite Congress. Soon the author received a reply from him, in which he expressed that he would like to communicate with those who have different points of views for in-depth discussion. With his constant encouragement and assistance, the senior author wrote two short articles entitled "Myrmekite formed by Ca- and Na-metasomatism of K-feldspar" (2002) and "Nibble metasomatic K-feldspathization" (2003) which were published on his website (http://www.csun.edu/~vcgeo005/ Myrmekite and Metasomatic Granite, ISSN 1526-5757). In 2006, the senior author accompanied by his son, Shenwen

Rong, visited Prof. Collins. With his warm invitation and guidance we made an on-the-spot investigation to Temecula and observed interesting thin sections. Afterwards, the senior author wrote the initial version of the book, i.e., a paper entitled "Two patterns of monomineral replacement in granites," which was published on his website in 2009 with his careful editorial handling. The above three papers were transformed by him into PDF documents (Nr45Rong1.pdf, Nr46Rong2.pdf and Nr55Rong3.pdf) in 2012.

The English version of this book was also attentively revised and carefully corrected by Prof. Lorence Collins. The authors would like to extend their special sincere thanks to Prof. Lorence Collins for his friendly assistance and frank communication.

The authors thank Academician Qihan Shen, Professors Zhendong You, and Letian Du for going over the manuscript and writing forewords for the book. The authors are indebted to Academicians Qihan Shen, Mingguo Zhai, and Xuanxue Mo for writing recommendation letters to apply for publication fund. The authors extend their heartfelt thanks to the publisher Springer-Verlag Berlin Heidelberg for the publication of English version. Much appreciation also goes to Peng Han, chief director, Yun Wang, editor of Geological Division of Science Press of China and Lisa Fan, editor of Springer for their active operation and hard work.

Contents

Two Major Patterns of Single Mineral Replacement in Granites

According to the authors' observation based on similarities or differences of crystallographic orientation of replacive and replaced minerals, metasomatic phenomena that occur in single minerals in granites can be basically classified into two major patterns:

(1) Hetero-oriented replacement or nibble replacement.
(2) Co-orientated replacement or topotaxial replacement.

Minerals of different kinds have different crystallographic lattices. So, metasomatic processes among them must be only of hetero-orientation, such as quartz or muscovite replaces K-feldspar, etc.

Minerals of the same or similar kinds, such as feldspar minerals and mica minerals, have the same or similar crystallographic lattices. So metasomatic processes between them should be divided into hetero-oriented metasomatism and co-oriented metasomatism.

1.1 Hetero-Oriented or Nibble Replacement Pattern

Hetero-oriented replacement occurs at the boundary between two mineral grains with different orientations. In other words, the crystallographic lattice orientations of replacive and replaced minerals must be different or discordant. The replacive mineral selects the mineral of the same or similar kind in rock as crystallographic center (nucleus or basis), metasomatically growing toward the adjacent mineral it replaces, as if nibbly eating it. The crystallographic orientation of the replacive mineral is consistent with the mineral of the same or similar kind upon which it leans and different from that of the mineral which it replaces. The unreplaced part of the replaced mineral survives in the replacive mineral without changing its crystallographic orientation.

No nibble replacement is possible along the boundary between two minerals (plagioclase and K-feldspar, for example) where they have the same and parallel crystallographic orientation.

Hetero-oriented replacement phenomena usually observed in granites are albitization, K-feldspathization, muscovitization, quartzification, as well as beryllization, calcitization and pyritization.

An impurity in the rock could serve as a crystal nucleus only when no identical or similar mineral is present and replacement should occur.

1.1.1 Hetero-Oriented Albitization

Hetero-oriented albitization in granites, especially rich in alkali, quite often appears at grain boundaries between plagioclase and K-feldspar as well as two K-feldspars.

1.1.1.1 Hetero-Oriented Albitization at Contact of Plagioclase with K-feldspar

A K-feldspar megacryst commonly contains inclusions of plagioclase. Albite rim (named 'clear rim' by Phemister 1926) often surrounds

© Science Press and Springer Science+Business Media Singapore 2016
J. Rong and F. Wang, *Metasomatic Textures in Granites*, Springer Mineralogy,
DOI 10.1007/978-981-10-0666-1_1

the plagioclase inclusions in K-feldspar. The width of albite rim is usually <0.1 mm, but can be as much as 0.3–0.6 mm in granites rich in alkali, alumina and silica. "There is usually a sharp and smooth contact from the main plagioclase (typically oligoclase) to the albite rim, whereas the albite-K-feldspar boundary is irregular. Some rims contain thin spindles of quartz oriented quasi-normal to the surface" (Smith 1974).

"Clear rim" occurs only at contacts of plagioclase with differently oriented K-feldspar (Figs. 1.1, 1.2 and 1.4), but is absent at the border between plagioclase with co-oriented K-feldspar (Pl$_1$ in Fig. 1.5) or between two plagioclases (Figs. 1.1 and 1.4) or between plagioclase (or K-feldspar) and quartz (Figs. 1.2 and 1.3).

Plagioclase itself may have a clear border due to the less calcium content (border part of

Fig. 1.1 (+). Clear rims (Ab$_2'$, Ab$_3'$) at contacts around plagioclase inclusions (Pl$_2$, Pl$_3$) in K-feldspar K$_1$. The growth of albite Ab$_2'$ is blocked at contact with another plagioclase Pl$_1$ and perthitic albite Ab$_1$. (−)—Plane polarized light; (+)—Crossed polars; (+)Q—Crossed polars plus quartz plate

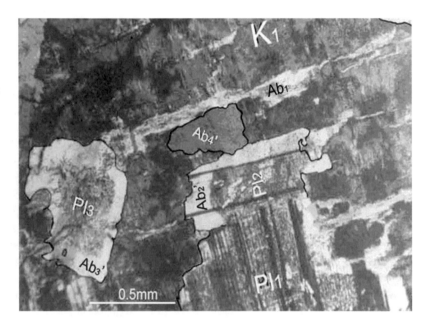

Fig. 1.2 (+). Albite rim is present at contact of Pl$_1$ with K, and absent at contact of Pl$_2$ with quartz Q. (−)—Plane polarized light; (+)—Crossed polars; (+)Q—Crossed polars plus quartz plate

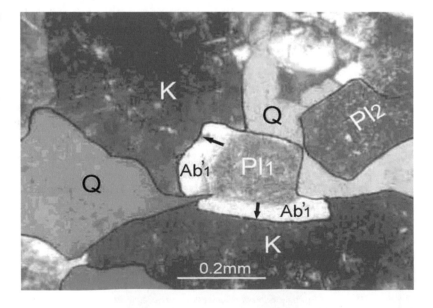

Fig. 1.3 (+). Clear rim Ab′ occurs also at the border of unsericitized Pl (An17) with K. (−)—Plane polarized light; (+)—Crossed polars; (+)Q—Crossed polars plus quartz plate

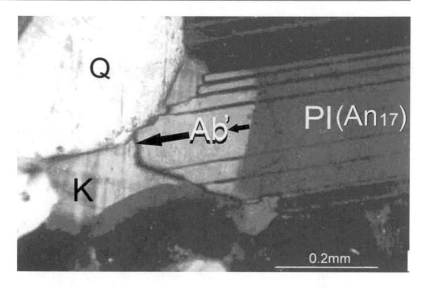

Fig. 1.4 A(+); B(+)Q. Clear rim appears at the border of plagioclases with differently oriented K. No clear rim appears at contact between two plagioclases. (−)—Plane polarized light; (+)—Crossed polars; (+)Q—Crossed polars plus quartz plate

(a)   (b)

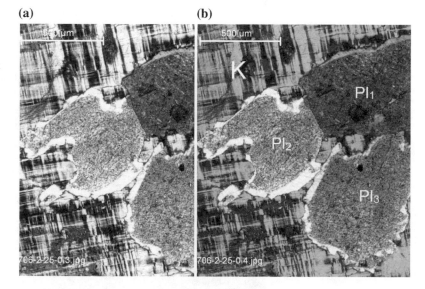

plagioclase in Fig. 1.9), but it is not a real "clear albite rim."

The development of "clear rim" is always blocked by perthitic albite (Fig. 1.1).

Under crossed polars plus quartz plate with careful observation, small relicts of perthitic albite can sometimes be noticed in clear rim, especially when the rim is thicker and the perthitic albite lamellae are more developed (Figs. 1.6, 1.7, 1.8, and 1.9).

The authors believe that a clear rim is formed from K-feldspar hetero-orientedly replaced by albite.

1.1.1.2 Hetero-Oriented Albitization Between Two K-Feldspar Grains

The albite grains occurring at grain boundaries between two K-feldspars (K_1 and K_2) with different orientation are called intergranular albite.

Fig. 1.5 A(+); B(+)Q. Clear rim appears at the border of host K-feldspar K and plagioclase Pl₂ with different orientation. No clear rim exists at the border of Pl₁ with K, when the enclosed Pl₁ has the same orientation as the host K

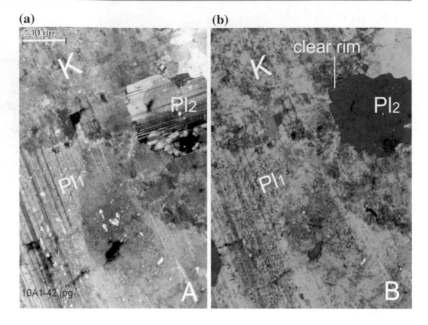

Fig. 1.6 (+)Q. A relict of perthitic Ab in clear rim Ab₁′. Naqin granite, Taishan County, Guangdong Province

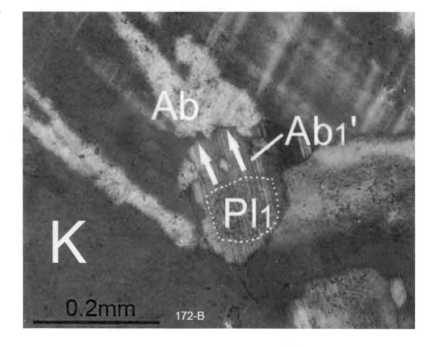

The intergranular albite grains can usually be divided into two rows (Ab₁′, Ab₂′) (Figs. 1.10, 1.11, 1.12, 1.13, and 1.14) when a quartz plate is inserted under cross-polarized light. Each row (generally <0.2 mm wide) has an approximate optical orientation as that of the perthitic albite of the K-feldspar on the opposite side of the contact and has a different orientation from that of the front K-feldspar that it replaces. One row might grow quite well, while the other might develop poorly (Fig. 1.13). It would be difficult to determine accurately the boundary between perthitic albite Ab₁ and intergranular albite Ab₁′ where they directly contact each other because

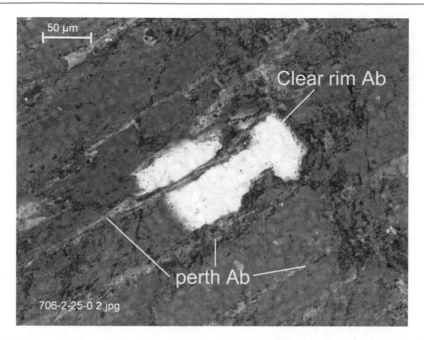

Fig. 1.7 (+)Q. It seems that the perthitic albite penetrated a plagioclase inclusion. In fact, the latter is a clear albite rim which replaced K-feldspar with a survival perthitic albite remained. Jiling granite, Gansu Province

Fig. 1.8 (+)Q. Intensely replacive albite (0.8 mm thick) toward K-feldspar. A—The isolated relicts of perthitic albite Ab₁ remain in the replacive albite Ab′. B—Local magnification of frame in A. Ab′ contains tiny myrmekitic quartz and several relics of Ab₁. Small myrmekitic quartz grains are noticed in Ab′. Naqin leucogranite, Taishan County, Guangdong Province

they have the same orientation and similar composition (Fig. 1.12).

Both clear albite rim and intergranular albite are situated at contacts with K-feldspars. Their thicknesses in a given rock are similar (Fig. 1.10), although intergranular albite has generally less continuity than that of clear albite rim.

Fig. 1.9 (+)Q. Generally, clear rim albite absents at contact of plagioclase with primary quartz. The unsericitized clear border of plagioclase contacting at quartz Q_1Q_2 is still plagioclase rather than clear rim albite

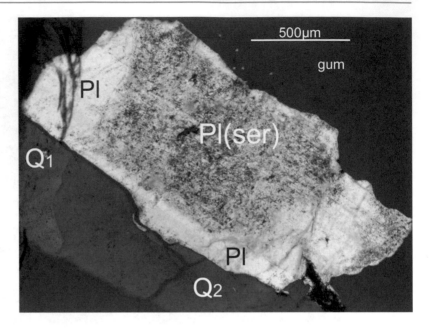

Fig. 1.10 (+)Q. Swapped rows of intergranular albite. Each row has the same optical orientation as the perthitic albite of the opposite K-feldspar. The thickness of swapped rows of albite is closely equivalent to that of clear rim around Pl_3. Naqin leucogranite

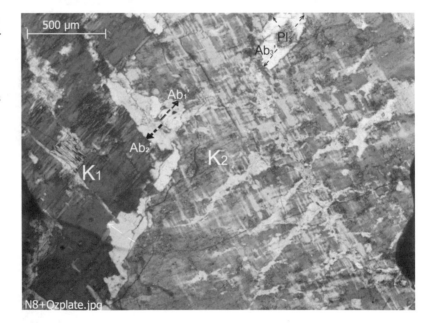

There are several explanations for the origin of clear albite rims and grain boundary albite grains:

(1) Unmixing (exsolution) from adjacent K-feldspar (Phemister 1926; Tuttle 1952; Voll 1960; Ramberg 1962; Phillips 1964; Hall 1966; Carstens 1967; Haapala 1997); Ramberg argued that the albite rim was generated by exsolution from the host crystal

K-feldspar. Some of the albite molecule held in solid solution exsolves inside the K-feldspar grain to form perthite. Other albite migrates by diffusion to nucleate on a plagioclase grain, or on an adjoining K-feldspar grain. Therefore, there are no rims at the boundaries of plagioclase-plagioclase, plagioclase-quartz or plagioclase-biotite.

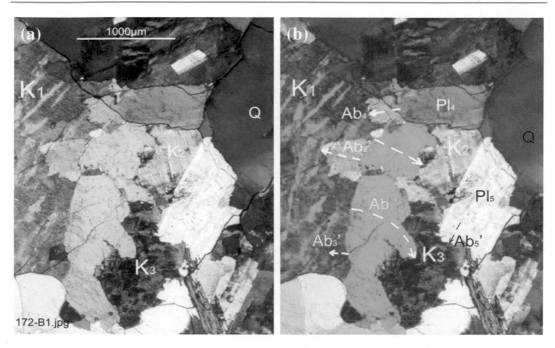

Fig. 1.11 A(+); B(+)Q. Swapped rows of intergranular albite (Ab$_1$′ Ab$_2$′ Ab$_3$′) and clear rim albite (Ab$_4$′ Ab$_5$′) have comparable width. Naqin leucogranite, Taishan County, Guangdong Province

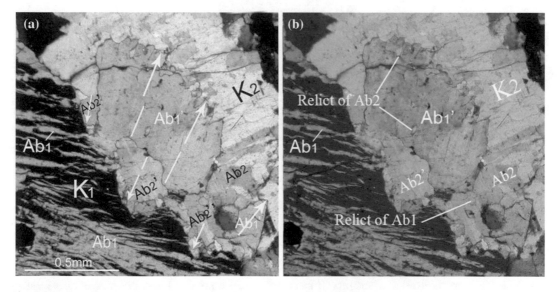

Fig. 1.12 A(+); B(+)Q. Swapped albite rims with relics of perthitic albite between K$_1$ and K$_2$. The contact between Ab$_1$′ with the perthitic albite Ab$_1$ is blured

Ramberg postulated that swapped rims of albite were attributed to the grain boundary diffusion migration which reduces surface energy, resulting in the formation of an albite rim on the neighboring K-feldspar across incoherent boundaries.

Fig. 1.13 (+)Q. The left row of albite Ab$_2$' contains relics of perthitic albite Ab$_1$ and tiny vermicular quartz

Fig. 1.14 A(+); B(+)Q. Relicts of perthitic albite Ab$_2$ remains in the replacive albite Ab$_1$'. Zhaojiayao granite, Xuanhua County, Hebei Province

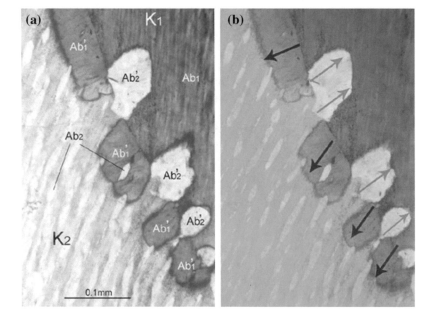

(2) Late stage crystallization of magma (Rogers 1961; Peng 1970; Hibbard 1995);

They believed that the irregular development of intergranular albite rims and the presence of fine rods of quartz inside the albite rims (other than perthitic albite which has no vermicular quartz) are inconsistent with exsolution from K-feldspar. So they are caused by the crystallization of interstitial residual melt and the accompanying replacement of K-feldspar (growth onto adjoining K-feldspar). Peng even considered the fact that none of rim albite at contacts with quartz was attributed to the quartz as the last crystallizing mineral from the melt (after the clear rim was formed).

(3) The sericite in the rim of sericitized plagioclase is dissolved away by sodic
hydrothermal solution (Cheng 1942; Cheng
et al. 1963);
Cheng supposed that plagioclase (mainly
oligoclase) was sericitized at first, and then its
rim was dissolved by Na-bearing hydrothermal fluid. The sericites on it had migrated
away, leading to the formation of clear rim
albite. Because no clear albite rim occurs at
the borders of plagioclase-plagioclase and
quartz-biotite, Cheng emphasized that the
presence of K-feldspar, i.e., contact with
K-feldspar, is probably in favor of dissolution of the rim side of sericitized oligoclase.
Cheng's hypothesis was approved by He
et al. (1980) and Wang (1989).

(4) Plagioclase is replaced by K-feldspar (Deer
1935; Schermerhorn 1956).

The authors of this book disagree with the
above four explanations and consider that they
may be formed by a similar or the same origin
since the width of clear albite rim and intergranular albite is comparable or similar in a given
rock (Figs. 1.10 and 1.11). Their genesis might
be found out by the following phenomena:

(1) Clear albite rim appears at contacts of plagioclase with K-feldspar no matter whether
the plagioclase is sericitized or unsericitized
(Fig. 1.3).

(2) Very fine vermicular quartz occurs in
intergranular or clear rim albite with An >3
(Figs. 1.8 and 1.13), even An > 1 (You
et al. 1996), but is generally absent in pure
albite rim (An $\approx$ 0).

(3) Clear rim is blocked where plagioclase
contacts quartz or another plagioclase (including perthitic albite) (Figs. 1.1, 1.2, 1.6
and 1.8).

(4) Tiny relicts of perthitic albite of outside
K-feldspar can sometimes be found in clear
rim or intergranular albite (Figs. 1.6, 1.8
and 1.12). The relicts strictly maintain the
primary orientation, although they are finer
and clearer. Similar but coarser relicts of
perthitic albite have also been found associated with clear albite where
Na-metasomatism has modified K-feldspar

in the Lyon Mountain granite gneiss in New
York (Collins 1997).

Although the continuity of intergranular albite
is less than that of the clear rim, the phenomena
are identical, indicating that they are of the same
origin (hetero-oriented replacement). Intergranular albite is divided into two rows called swapped
albite rows (Voll 1960).

The relation between perthitic albite and clear
rim, as well as intergranular albite, is usually hard
to determine because of the following reasons:

(1) The chance of contact between them is rare
due to their limited development;

(2) No relict of perthitic albite in clear rim or
intergranular albite can be seen;

(3) Perthitic albite in form of a wedge or comb
penetrates the clear rim (Fig. 1.8) or intergranular albite (Fig. 1.13), and in some
places even as a veinlet that penetrates the
intergranular albite (Fig. 1.7) as if the perthitic albite replaces the intergranular albite
(also see Figs. 428, 429 in the book "Atlas
of the textural patterns of granites, gneisses
and associated rock types" by Augustithis
1973). The bridge-like veinlets are, in fact,
the relicts of perthitic albite that survived
albite replacement and remained in the
replacive albite;

(4) A quartz plate is not inserted under
cross-polarized light.

Although it is almost impossible for albite to
hetero-orientedly replace an entire plagioclase
crystal, the replacive albite may still replace tiny
perthitic albite. Moreover, the presence of tiny
relicts of perthitic albite in the replacive albite is
easily missed if the quartz plate is not inserted.

Microprobe analyses of perthitic albite and
swapped metasomatic albite are listed in
Table 1.1.

(Microprobe analyses in the book are all
conducted with spectrum analyses, accelerating
voltage 20 kV, beam current 1×10^{-8} A by
microprobe analyser JXA-8100 in Beijing
Research Institute of Uranium Geology).

Table 1.1 indicates that the compositions of
swapped albites range from nearly pure albite
(Fig. 1.13) to sodic oligoclase (Fig. 1.14). In most
places, tiny vermicular quartz may occur in nibbly

Table 1.1 Microprobe analyses of perthitic albite Ab_1 and swapped metasomatic albite Ab_1'

	Mineral	Mean	SiO_2	Al_2O_3	CaO	Na_2O	K_2O	Total	An
N_4 (Fig. 1.13)	Perth.Ab_1	5	68.13	19.40	0.26	11.75	0.09	99.78	1.2
N_4	Swap.Ab_1'	3	67.81	19.89	0.31	11.76	0.16	99.94	1.4
CL_3 (Fig. 1.14)	Perth.Ab_1(Olig)	3	67.58	20.59	3.15	8.32	0.15	99.94	17.4
CL_3	Swap.Ab_1'(Olig)	2	67.59	19.8	2.44	9.95	0.09	99.96	12

N_4 from Naqin granite, Taishan County, Guangdong Province. CL_3 from Zhaojiayao granite, Xuanhua County, Hebei Province

metasomatic albite with An value >3–5. Nevertheless, no vermicular quartz appears in perthitic oligoclase in sample CL_3, even its An value is as high as 17.4.

It shows that the so-called perthitic "albite" may not be albite, but acid oligoclase as well and the swapped metasomatic "albite" may be oligoclase too. In addition, the composition of the swapped albite is seemingly positively related with that of the perthitic albite.

1.1.2 Hetero-Oriented K-feldspathization

The minerals replaced hetero-orientedly by K-feldspathization are mainly plagioclase and K-feldspar.

1.1.2.1 K-feldspar Hetero-Orientedly Replaces Plagioclase

Plagioclase crystals are often enclosed by K-feldspar.

Where hetero-oriented K-feldspathization occurs, the grain boundaries of plagioclases with K-feldspars become complicated and tortuous. Some remnants of plagioclase are found at the borders of the K-feldspar (Figs. 1.15 and 1.16).

This phenomenon is always treated as an important evidence for intense K-feldspathization, and therefore, the megacryst of K-feldspar has a porphyroblast origin. Surely, the plagioclase is actually replaced by K-feldspar, but the real replacive K-feldspar that replaced the plagioclase is mainly situated in the limited area around the relicts of plagioclase nearby. The major part of the K-feldspar away from the limited area may not be

Fig. 1.15 A(+); B(+)Q. Plagioclase Pl is replaced by the K_1' which lacks perthitic albite. Da-ao granite, Yangjiang County, Guangdong Province

(a)   **(b)**

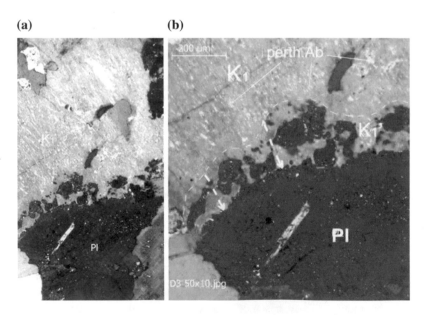

Fig. 1.16 A(+); B(+)Q. Pl
is replaced by K'.
Compared with K, the K'
lacks evidently perthitic
albite. Porphyritic biotite
granite, Zhuguangshan
batholith, Guangdong
Province

(a)   **(b)**

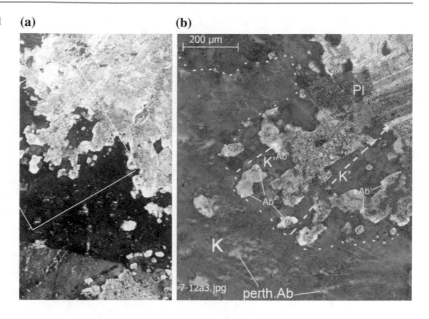

Fig. 1.17 (+)Q. Several
plagioclase grains are
hetero-orientedly replaced
by K-feldspathization.
Only the co-orientedly
enclosed Pl₁ in K survives.
Da-ao granite, Yangjiang
County, Guangdong
Province

metasomatic. It is hard to judge whether the whole uniform K-feldspar has a metasomatic origin.

Figures 1.15 and 1.16 show that the newly formed (metasomatic) K-feldspar, penetrating into the relict areas, is comparatively pure and contains less perthitic albite than that in the outside surrounding K-feldspar. The authors consider that the latter is the preexisting K-feldspar (probably primary) and the former has a metasomatic origin. Nevertheless, if the pre-existing K-feldspar (K_3 in Figs. 1.17 and 1.18) lacks perthitic albite, too, there is no obvious difference between primary K_3 and metasomatic K_3' under the microscope, and additionally, no evident divergence in chemical component (Table 1.2). As a result, it is easy to assume that the whole K-feldspar has a metasomatic origin (Fig. 1.19).

Fig. 1.18 (+). Pl$_1$ is partly replaced by K$_3$'. Both K$_3$ and K$_3$' lack perthitic albite. Da-ao granite, Yangjiang County

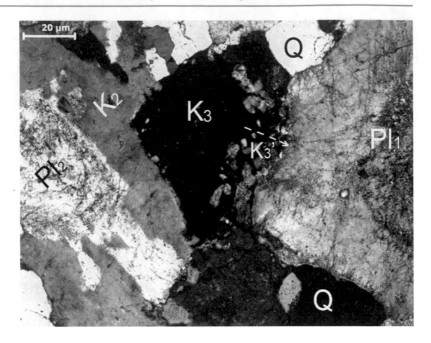

Table 2 Microprobe analyses of primary K-feldspar K$_3$ and metasomatic K-feldspar K$_3$' replacing plagioclase

Mineral	Mean	SiO$_2$	Al$_2$O$_3$	CaO	Na$_2$O	K$_2$O	Total
K$_3$ (Fig. 1.18)	5	66.07	17.44	0.004	0.38	16.06	99.95
K$_3$' (Fig. 1.18)	5	66.46	17.24	0.01	0.29	16.09	100.1

Sample from Da-ao granite

Fig. 1.19 (+)Q. K-feldspathization K$_3$' occurs at the border of K$_3$ with the differently oriented Pl$_1$. No K-feldspathization occurs at contacts of Pl$_2$ with K$_2$ when they have the same orientation

Table 1.2 shows similar compositions between primary K-feldspar K_3 and metasomatic K-feldspar K_3' (replacing plagioclase).

However, there is a subtle difference in the cathodoluminescent image (Fig. 1.20), i.e., Russian blue triangle tone appears on K_3 (free of relicts) but is absent on K_3' (containing relicts).

Note that the orientation of the replaced plagioclase inclusions are obviously different from that of the surrounding K-feldspar. The plagioclase (Pl_1 in Fig. 1.17 and Pl_2 in Fig. 1.18) having the same orientation as the K-feldspar nearby may remain unchanged during hetero-oriented K-feldspathization. Therefore, hetero-oriented K-feldspathization affects only those plagioclases with different orientation from that of the surrounding K-feldspar. The complete preservation of some plagioclases from replacement is not because of their strong structure, i.e., less likely to be fractured, as deduced by Collins and Collins (2002a), but just because their parallel lattice (epitaxial) boundaries with K-feldspar are strictly sealed, preventing introduction of gas or liquid that initiate replacement.

1.1.2.2 New K-feldspar Nibbly Replaces Old K-feldspar

The appearance of hetero-oriented (nibble) K-feldspathization, i.e., old K-feldspar being replaced by newly formed K-feldspar, occurring irregularly in the K-rich quartz syenite or monzonite rocks is quite unique at first glance under cross-polarized light. The K-feldspathization initially found by the senior author in Huangnitian hornblende quartz syenite, Yangjiang County, Guangdong Province, developed so intensely that the volume of newly formed K-feldspar makes up 12–35 % of the whole rock.

Generally, the ordinary K-feldspars are semi-euhedral or granular in form and have minor content of perthitic lamellae. However, among them appears a new kind of K-feldspar which is characterized by the presence of tiny isolated relics of perthitic lamellae of the surrounding ordinary K-feldspar (Figs. 1.21 and 1.22) and of special vermicular plagioclase (<10–50 μm wide and 50–200 μm long) radially pointing toward the border (Figs. 1.21 and 1.23). The disseminated relics of mainly perthitic albite

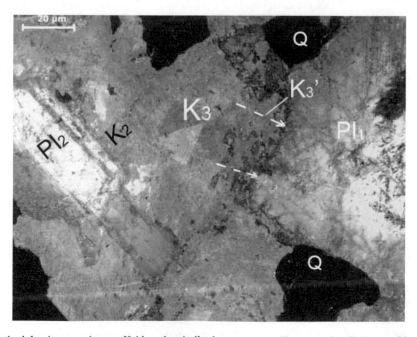

Fig. 1.20 Cathodoluminescent image. K_3' has the similar brown tone as K_3 except less *light gray blue*

Fig. 1.21 A(+); B(+)Q. K-feldspar is complicated. Newly formed K_2' contains relicts of perthitic Ab_1 of K_1 and vermicular albite. Huangnitian quartz syenite, Yangjiang County, Guangdong

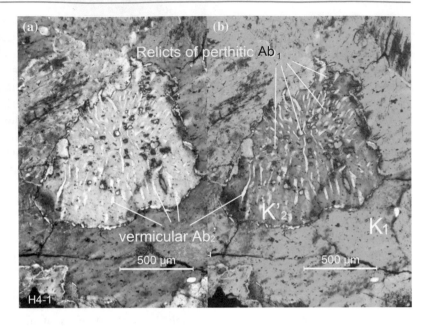

Fig. 1.22 (+)Q. K_2' K_3' replaced K_1 only. Pl and Q survived K-feldspathization. Afterwards, week swapped albite rows took place along borders of K_1 and K_2', K_1 and K_3'. Huangnitian quartz syenite

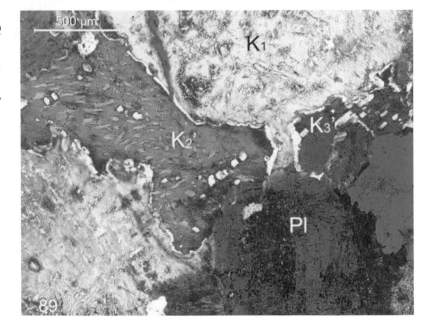

maintain their original orientation as that of the surrounding K-feldspar and the orientation of the vermicular plagioclase is the same as that of the host new K-feldspar. The orientation of the new K-feldspar is surely different from that of the surrounding old K-feldspar, where the tiny relics and vermicular plagioclase are absent. This strange phenomenon is obviously and amazingly noticed under microscopic observation when a quartz plate is inserted.

The fact that the K-feldspar contains the above remnants of perthitic albite of the surrounding K-feldspar is an important unique evidence showing that the K-feldspar is formed by hetero-oriented metasomatism. Accordingly, the K-feldspars without relics of perthitic albite in the same thin

Fig. 1.23 (+)Q. Replacive K₃′ bearing vermicular albite has the same orientation as that of the primary K₃. Weak albitization (swapped) occurs along the border of the two K-feldspars. Huangnitian quartz syenite

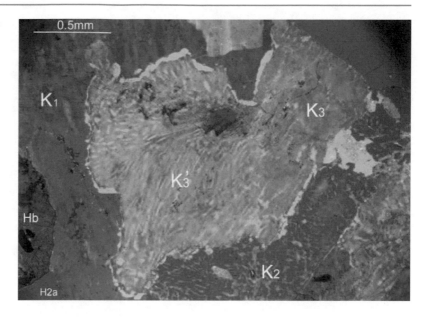

Fig. 1.24 (+)Q. K′ replaces K with whole Pl and part of K remained. K′ contains perthitic Ab′ instead of vermicular albite. Weiya hornblende quartz syenite, Hami County, Xinjiang Autonomous Region

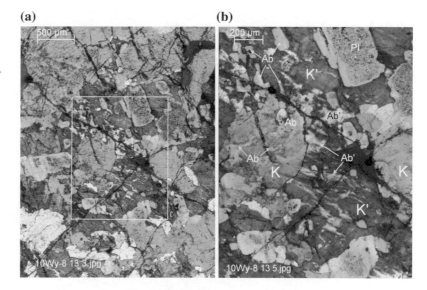

section may be judged as of primary origin. No co-oriented metasomatic K-feldspathization in thin sections has been found.

The plagioclase vermicules in new K-feldspar may serve as additional evidence that the latter is metasomatic in origin, especially in case of rare presence of the relics of perthitic albite from the replaced K-feldspar (Fig. 1.23).

Nevertheless, the radially distributed vermicular plagioclase may not necessarily exist in the replacive K-feldspar (Figs. 1.24, 1.26, 1.27).

The orientation of metasomatic K-feldspar ($K_x′$) is strictly coincides with the back K-feldspar (K_x) on which it epitaxially grew (Figs. 1.23, 1.25, and 1.26).

Because the size of replacive K-feldspar may be as large as 0.5–2 mm, the primary K-feldspar on which the replacive K-feldspar nucleated naturally may not appear simultaneously in one thin section (Figs. 1.21 and 1.22).

The swapped pieces of hetero-oriented metasomatic K-feldspar can still fortunately be

Fig. 1.25 (+)Q. Intense swapped K-feldspathization $K_1'K_2'$ between two primary K-feldspars K_1K_2. Neither relics of perthitic albite nor vermicular albite are contained in the primary K-feldspars K_1K_2. Huangnitian quartz syenite

Fig. 1.26 (+)Q. Relicts of K_1K_2 are enclosed in swapped $K_1'K_2'$ which have no clear myrmekitic Ab. Huangnitian quartz syenite

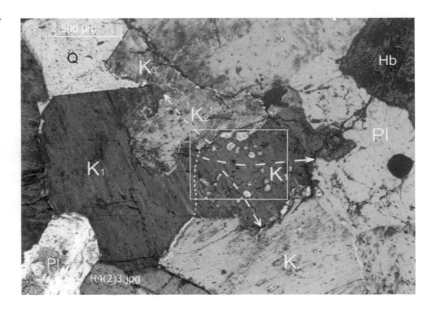

observed at grain boundaries between two differently oriented primary K-feldspar crystals (K_1, K_2) (Figs. 1.25, 1.26 and 1.29) as swapped rows of albite occur at contacts between two K-feldspars. The replacive K-feldspar (K_1') with the same orientation of K_1 on which the K_1' nucleated replaces the differently oriented K-feldspar (K_2), whereas the replacive K-feldspar (K_2') with the same orientation as K_2 on which the K_2' is attached replaces the

differently oriented K_1. Both replacive K-feldspars, K_1' and K_2', are distinguished from the primary K-feldspar, K_1 and K_2, respectively, by the presence of residual inclusions of foreign perthitic albite (Ab_1, Ab_2), even old K-feldspar (Figs. 1.24 and 1.30) and vermicular authigenic perthitic albite (Ab_1', Ab_2') (Fig. 1.31).

The major replaced minerals in K-feldspathization are K-feldspars. Some perthitic albite can also be replaced. The mineral on

Fig. 1.27 (+)Q.
Enlargement of frame in
Fig. 1.26. Relicts of
perthitic plagioclase hold
the same orientation.
A weak albitization
(swapped $Ab_1''Ab_2''$) is
formed after
K-feldspathization

Fig. 1.28 (+)Q.
Enlargement of frame in
Fig. 1.27. A big relict of
perthitic albite is divided
into core ($An_{12.4}$) and rim
($An_{0.8}$)

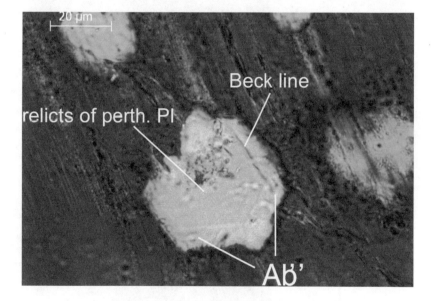

which the preferential epitaxial growth of new
K-feldspars leans is only old K-feldspar.

K-feldspathization stops or is blocked at contact
with whole plagioclase (Figs. 1.22, 1.24 and 1.29),
hornblende (Fig. 1.32), pyroxene (Fig. 1.29), and
quartz (Fig. 1.32), indicating that they are stable
during such intense K-feldspathization.

After K-feldspathization a weak albitizaton
causes the swapped albite <20 μm wide formed at
contacts between two differently oriented
K-feldspars (old and new) (Figs. 1.21, 1.23 and
1.27). In some places remnants of perthitic albite

may be wider than the width of perthitic albite in the
replaced K-feldspar. Careful observations of an
isolated wider albite remnant at high magnification
and with a reduced diaphragm show that a Becke
line divides the remnant into two parts: an inner
nucleus and an outer rim (Fig. 1.28). The inner
nucleus has a slightly higher refractive index and
lower birefringence, indicating it has a comparatively higher An content than that of the outer rim,
although there is always no abrupt boundary
between them under normal cross-polarized light. It
shows that the inner nucleus is of metasomatic

Fig. 1.29 (+)Q.
K-feldspathization (K_1',
K_2') at contact between two
primary K-feldspars (K_1,
K_2). Plagioclase Pl and
pyroxene Px survive.
Hengling monzonite,
zhuguangshan batholith,
Guangdong Province

Fig. 1.30 (+)Q.
Enlargement of Fig. 1.29.
Relicts of K-feldspar K_1
and plagioclase Pl remain
in the replacive K-feldspar
K_2' containing tiny fine
vermicular albites

residue of a perthitic lamella of replaced K-feldspar, whereas the outer albite rim around the residue is formed by nibble albite replacement toward the newly formed K-feldspar. This phenomenon is consistent with that of swapped albite rows. Both are of nibble albite replacement. The authors are much puzzled by the euhedral contour of the inner core (Fig. 1.28). The weak hetero-oriented albitization formed after K-feldspathization appears widely in hornblende quartz syenite (Yangjiang County, Guangdong Province and Hami County, Xinjiang Autonomous Region), while absents in more basic biotite pyroxene monzonite (Fig. 1.29).

Figures 1.31 and 1.32 show that the metasomatic K-feldspars (K_1', K_2', K_3') nibbly replace the adjacent K-feldspars (K_2, K_1, K_3) on their

Fig. 1.31 (+). Replacive K-feldspars at grain boundaries among primary K-feldspars $K_1K_2K_3$. Huangnitian hornblende quartz syenite, Yangjiang County, Guangdong Province

Fig. 1.32 (+)Q. Primary K-feldspars $K_1K_2K_3$ and replacive $K_1'K_2'K_3'K_4'$. Each forward replacive K-feldspar ($K_1'K_2'K_3'$) has the same orientation as the K-feldspar situated behind ($K_1K_2K_3$), respectively. Hornblende Hb and quartz Q survive

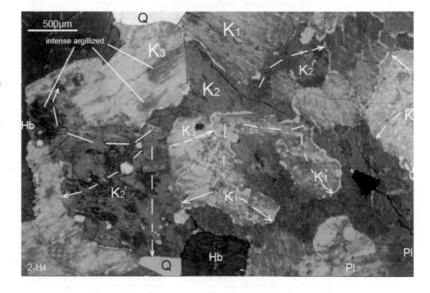

boundaries. K_1' replaces K_2 downwards, K_2' replaces K_3 leftwards and K_1 rightwards, and K_3' replaces K_2 rightwards, respectively.

The cathodoluminescent image (Fig. 1.33) indicates that the primary K-feldspar has pale pinkish purple luminescence, that the replacive K-feldspar in many places has bright sky-blue luminescence. But, part of the bright sky-blue color also extends to the adjacent primary K-feldspar.

All the intense argillized parts of K-feldspar have dead-wood tan color. Both the pinkish purple (for primary K-feldspar) and bright sky-blue (for replacive K-feldspar) disappear at the intensely argillized parts of the K-feldspar.

The orientation of the metasomatic K-feldspar (Kx′) is different from that of the replaced K-feldspar, but coincides with the adjacent K-feldspar (Kx) on which it epitaxially grows.

Microprobe analyses of primary K-feldspar K, metasomatic K-feldspar K′, perthitic albite Ab in K, and myrmekitic albite Ab′ in K′ are listed in Table 1.3.

The nibbly replacive K-feldspar has higher Na_2O, CaO, and lower K_2O contents in comparison with those components in the primary K-feldspar,

Fig. 1.33 Cathodoluminescent image. $K_1K_2K_3$ shows pale purple, while $K_1'K_2'K_3'K_4'$ presents bright *sky blue*. All the intense argillized K-feldspar displays wither tree brown. Pl shows light brown

Table 3 Microprobe analyses of primary K-feldspar K, replacive K-feldspar K', perthitic albite Ab in K and myrmekitic albite Ab' in K'

Mineral (Fig. 1.31)	Mean	SiO_2	Al_2O_3	CaO	Na_2O	K_2O	Total	An
Primary K	15	65.39	17.85	0.04	0.82	15.8	99.9	–
Variation range	–	64.4–67	16.7–18.6	0–0.1	0.2–1.8	14.3–16.7	–	–
Replacive K'	16	65.61	18.1	0.09	2.19	13.7	99.7	–
Variation range	–	64.2–67.1	17–18.7	0–0.2	0.6–3.6	11–16.3	–	–
Perth.Ab in K	10	67.6	19.78	0.98	10.96	0.29	99.6	4.71
Variation range	–	66.3–69.3	18–20.8	0.2–1.9	10.2–11.4	0.1–0.4	–	1–10
MyrmAb' in K'	15	66.18	20.4	2.22	10.67	0.24	99.89	10.31
Variation range	–	63.2–67.6	18.5–21.8	0.8–4.9	9.8–12	0.16–0.36	–	2–21
Swap.replac.Ab' (H_4)	3	68.9	18.08	0.55	11.91	0.08	99.52	2.5
Resid. core of perth. Pl (H_4)	–	64.99	21.2	4.25	9.12	0.25	99.81	20.5
New rim Pl (H_4)	3	67.19	19.78	2.47	10.47	0.25	100.1	11.5
Resid. core of perth. Pl (H_2)	–	65.69	21.03	2.52	9.84	0.28	99.36	12.4
New rim Ab (H_2)	–	68.19	19.11	0.15	11.08	0.14	98.62	0.83

Sample from Huangnitian quartz syenite

but no evident difference exists in the variation range.

The myrmekitic albite in newly formed K-feldspar has double CaO content (An values) on the average as compared with that of the perthitic albite in primary K-feldspar. The An value of the former is twice higher than that of the latter.

The composition of perthic plagioclase in primary K-feldspar and two bigger relic grains of perthitic plagioclase are measured by microprobe.

Fig. 1.34 An unusual perthitic K-feldspar, containing vermicular plagioclase lamellae, occurs in monzonitic rocks from the Maronia pluton in northern Greece. (Work done by Georgios Christofides) Reported by Collins (1998) (With kind permission from Collins)

The An value of perthitic plagioclase varied in a wide range (0.6–9.7) with a mean of 4.71. A bigger relict grain (H₄) has core of An 20.5 and rim of An 11.5, while the another relict grain (H₂) has core of An 12.4 and rim of An 0.83, respectively. The An value of the inner core of relicts is higher than that of perthitic albite. Generally speaking, the inner core of relicts represents the residual part of perthitic plagioclase, though the former's An value is higher than the latter's, according to the limited data. While the outer rim is not residual, but is formed by a weak hetero-oriented albitization toward the surrounding new K-feldspar after K-feldspathization had ceased. The An value of newly formed albite-oligoclase is obviously lower than that of the original perthitic plagioclase.

A phenomenon that ordinary K-feldspar is invaded by unusual perthitic K-feldspar, containing vermicular plagioclase lamellae (Fig. 1.34) in monzonitic rocks from the Maronia pluton in northern Greece has been reported by Collins (1998). The initial material came from Georgios Christofides (Department of Mineralogy, Petrology, and Economic Geology, Aristotle University of Thessaloniki, 54006 Thessaloniki, Macedonia, Greece). The unusual K-feldspar is quite similar to the replacive K-feldspar mentioned above.

It is noted that vermicular albite occurs often in newly formed K-feldspar when it replaces primary K-feldspar [except in Weiya hornblende quartz syenite (Fig. 1.24)], while no vermicular albite occurs in the newly formed K-feldspar when it replaces primary plagioclase (Figs. 1.15, 1.16 and 1.18).

1.1.3   Hetero-Orientated Muscovitization

Hetero-oriented muscovitization may replace K-feldspar, biotite, and plagioclase. But, first we should deal with primary muscovite.

1.1.3.1   Primary Muscovite

Muscovite in granite was thought to be a typical epimagmatic mineral (Saavedra 1978; Speer 1984). It likely meant that muscovite in granite is metasomatic in origin. However, the experimental studies that the stability curve of muscovite (Yorder and Eugster 1955) and the equilibrium curve of muscovite plus quartz (Althaus 1970) cut diagonally the minimum melting curve of granite (Tuttle and Bowen 1958) in *PT* diagram at point of 700 °C, 1.5 kbar and 650 °C, 3.5 kbar, respectively, show that at the overlapping *PT* area of the above intersecting curves muscovite may crystallize as a primary mineral from a granitic magma (Fig. 1.35). Moreover, muscovite has been reported as a magmatic phase (small phenocryst) in Permian rhyolite in Germany (Schleicher and Lippolt 1981) and even in rhyolite melt inclusions (Webster and Duffield 1991). Therefore, it proved that muscovite can be formed as a primary (magmatic) phase in granite.

Primary muscovite is characterized as follows (Saavedra 1978; Speer 1984):

(1) Grain size comparable to other magmatic minerals or even coarser;
(2) Sharp grain boundaries (no relation with other minerals);
(3) Allomorphic, subhedral, or euhedral shape;
(4) Occurring in fresh unaltered plutonic rocks.

In addition, the authors add that the primary muscovite is comparatively clean, transparent

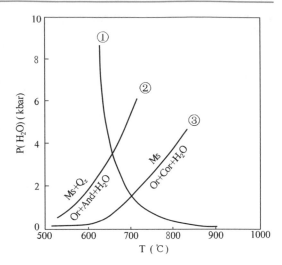

Fig. 1.35 Relation between stability curve of Ms and equilibrium curve of Ms+Qz with minimum melting curve of granite. ① Minimum melting curve of Granite (Tuttle and Bowen 1958); ② Equilibrium curve of Ms+Qz=Or +Al$_2$SiO$_5$+H$_2$O (Althaus 1970); ③ Stability curve of Muscovite Ms=Or+Al$_2$O$_3$+H$_2$O (Yoder and Eugster 1955)

and contains a few impurity and may enclose idiomorphic biotite either epitaxially (Figs. 1.36 and Fig. 1.37) or hetero-orientedly (Ms$_2$ in Fig. 1.37) and contacts quartz in many places as

Fig. 1.36 (+). Primary muscovite Ms epitaxially grows on idiomorphic biotite Bi. Huichang two-mica granite, Jiangxi Province

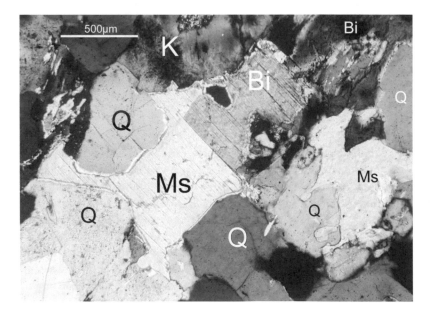

Fig. 1.37 (+). Primary muscovite Ms₁ Ms₂ enclosing and contacting euhedral biotite Bi and plagioclase Pl. Baimianshi two-mica granite, Huichang County, Jiangxi Province

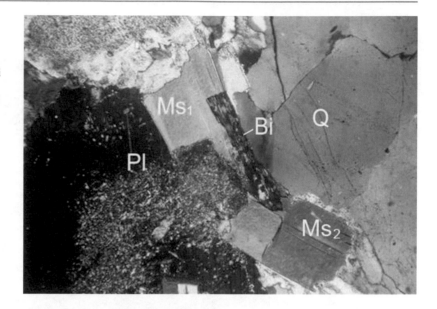

Fig. 1.38 (+). Interstitial primary muscovite Ms. Baimianshi two-mica granite, Huichang County, Jiangxi Province

a late stage crystallizing mineral. It is usually interstitial or fills other minerals (Fig. 1.38).

The grain boundary of primary muscovite should be sharp if secondary muscovitization absents. If the latter presents, the boundary contacting with quartz or plagioclase is still sharp (Fig. 1.39), while the boundary with K-feldspar may be unflattened or rugged due to muscovitization (Fig. 1.40). Secondary muscovitization usually occurs in Al-rich granites, especially in leucocratic peraluminous granites bearing primary muscovite, i.e., two-mica granites.

Many researchers considered that secondary muscovite can be divided into two kinds: late magmatic-post magmatic metasomatic muscovite (mainly replaces biotite) and hydrothermal flaky muscovite (mainly replaces plagioclase or along fractures).

The authors think that secondary metasomatic muscovite can be divided into two

Fig. 1.39 (+)Q. Euhedral K-feldspar K is replaced by Ms′ which has the same orientation as primary Ms outside the K nearby. Q survives. Baimianshi two-mica granite

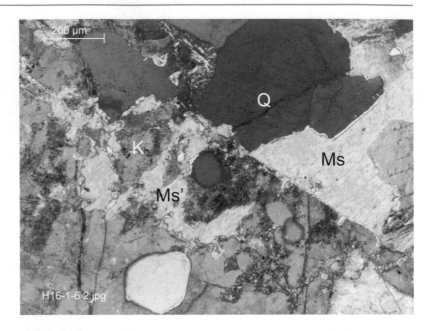

Fig. 1.40 (+). Microcline Mi is replaced by Ms′. The basal cleavage of Ms′ completely coincides with Ms. Pl and Q survive. Note the boundary of Ms with Pl, Q, is sharp, while the boundary of Ms with Mi is unflattened and rugged due to muscovitization

patterns: hetero-oriented and co-oriented. The co-oriented muscovitization replaces biotite only, which will be described in Sect. 1.2.1. The hetero-oriented metasomatic muscovite may replace mainly K-feldspar and also biotite, even replace the interior part of plagioclase as well.

1.1.3.2 Hetero-Oriented Muscovitization Replaces K-feldspar

K-feldspar can be easily replaced by muscovite. Hetero-oriented replacive muscovite grows often toward K-feldspar (Figs. 1.39 and 1.40). The thick blocky muscovite (Ms) bordering upon quartz (Q) outside the euhedral K-feldspar is primary, while the irregular branch-like muscovite (Ms') zigzag occurring in K-feldspar (K) is obviously formed by replacement. The replacive muscovite Ms' has strictly the same orientation as the primary muscovite Ms on which Ms' is attached. When K-feldspar is less euhedral or muscovitization is more intense, it is likely to treat the whole muscovite as metasomatic, since there is no differential boundary between primary and metasomatic muscovites in optical features.

1.1.3.3   Swapped Rims of Muscovite Between Two Biotite Grains

Hetero-oriented muscovitization replacing biotite may occur at the boundary of one biotite with another biotite (Fig. 1.41) or primary muscovite.

At the boundary between two leucocratic biotite (protolithionite) grains in Beihuan leucogranite, Yangjiang County, Guangdong Province, swapped rows (or rims) of small grains of muscovite nibbly replacing the two adjacent biotite can be observed (Fig. 1.42). Each row of

Fig. 1.41 A(−); B(+). Muscovite Ms_2' replacing Bi_1 between two biotite grains. Baimianshi two-mica granite

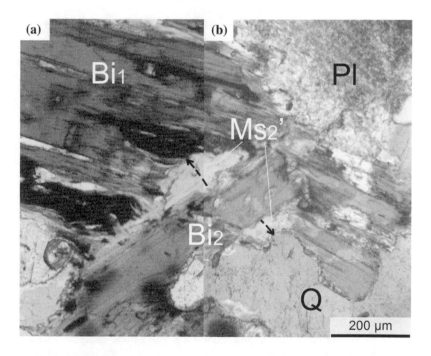

Fig. 1.42 A(−); B(+). Swapped rows of replacive muscovite (Ms_1', Ms_2') on border between two protolithionites (Bi_1, Bi_2). Beihuan leucogranite, Yangjiang County, Guangdong Province

Fig. 1.43 (+). Plagioclase
is transformed to albite
after sericitization and
muscovitization

muscovite has the same crystallographic orientation as the opposite biotite (according to the consistency of cleavage as well as the difference in pleochroism and interference color). It proves the real presence of hetero-oriented muscovitization replacing biotite. This phenomenon, however, is rarely present or seldom noticed due to the limited contact of two protolithionites and the tiny growth of replacive muscovite.

1.1.3.4   Sericitization, Muscovitization in Altered Plagioclase

Although no muscovitization happens on the border of plagioclase, sericitization (tiny flaky muscovitization) often (even always) occurs inside plagioclase. Many researchers treat sericitization of plagioclase as product of hydrothermal metasomatism in post-magmatic stage.

Plagioclase becomes muddy or turbid at the beginning of alteration, and then inside the muddy plagioclase a lot of tiny sericite, even fine muscovite are formed chaotically, which are independent of the muscovite or biotite outside the

plagioclase (Fig. 1.43). The occurrence and growth of sericitization is thought to be closely related with the appearance of micropores[1] inside plagioclase after scanning and transmission microprobe studies in recent 20–30 years. By reaction of hydrothermal fluid, micropores are much easier formed in plagioclase (especially oligo-andesine) than in K-feldspar.

Hydrothermal fluid may infiltrate micropores and decompose plagioclase, moving CaO away, leaving Na_2O, Al_2O_3, and SiO_2 behind. The K_2O and H_2O in fluid integrate with part of the latter (Al_2O_3 and SiO_2), forming sericite in micropores. Then sericite locally replaces plagioclase. Since no mica can serve as crystallizing center, the sericite has to grow on impurities or lattice defects. So, it also belongs to hetero-oriented replacement. Please refer to Sect. 2.5.

---

[1]The development of micropores is thought to be a major factor for the formation of co-oriented albitization. Nevertheless, micropores may also be present in K-feldspar but to a lesser extent. The genesis of micropores waits for further study.

1.1.3.5 Chemical Component of Muscovite of Various Geneses

The authors carried out a small amount of microprobe analysis for primary and metasomatic muscovite from two-mica granites in Jianxi Province, southern China compared with that in Southeastern California studied by Millers (Table 1.4). It looks like that the primary muscovite has relatively higher content of TiO_2, Al_2O_3, and Na_2O and lower FeO, MgO as compared with the metasomatic one. It was reported that the Ti content of primary muscovite in the leucogranites of the Massif Central, France, decreased systematically from core (0.45) to margin (0.2) (Monier and Robert 1986). Elevated Ti appears to be a fairly reliable indication of primary origin. The compositional zoning was explained by the rapid depletion of Ti in the magma rather than the subsolidus alteration processes. Most analytical studies on muscovite of different kinds show that TiO_2 content of primary muscovite is generally relatively higher than that of metasomatic one, while the contents of Mg, Fe, Na, Al, and Si are quite uncertain. In comparison with metasomatic muscovite, some researchers insist on relatively higher content of Fe, Si and lower of Na, Mg, and Al for primary muscovite (Borodina and Fershtater 1988; Wang et al. 2006), while others hold opposite conclusion (Sun et al. 2002). Recently the authors carried out microprobe analysis on the replacive muscovite Ms' hetero-orientedly replacing biotite (Fig. 1.41) in Baimianshi two-mica granite, Jiangxi Province.

The TiO_2 content (average of 16 measure data) of replacive muscovite reaches up to 1.24 %, double of the primary muscovite and the $\sum FeO$ and MgO content is also evidently higher than that of the primary one, indicating the inconformity with the above tendency. Zhang et al. (2010) considered that the chemical features of muscovite may not be of practical significance in distinguishing the genetic types of muscovite after they had studied Fucheng two-mica granite, Jiangxi Province.

The authors think, however, it is still necessary to accumulate the data before obtaining clear and definite conclusion whether there is a common tendency for composition disparity between the two genesis muscovite in different granites.

"Sericitization" of plagioclase is just a general term. Sericite (mainly illite) is usually accompanied by clay-like mineral, such as montmorillonite, mixed layer mineral, and kaolinite in various proportions. It is difficult to separate and

Table 4 Microprobe analysis of various muscovites (Ms-primary, Ms' K-replacing K, Ms' Bi-hetero-orientedly replacing Bi, Ms″-sericite in plagioclase)

Mineral	Sample number	SiO_2	TiO_2	Al_2O_3	$\sum FeO$	MgO	Na_2O	K_2O	Total
Ms (325-6)	3	46.2	0.76	31.69	3.1	1.08	0.6	10.71	94.17
Ms' K (325-6)	3	46.86	0.38	29.73	4.6	1.43	0.36	11	94.8
Ms (3075-2)	6	46.83	0.59	32.37	1.38	0.63	0.53	10.88	93.2
Ms' K (3075-2)	6	46.77	0.09	32.23	1.53	0.74	0.33	11.1	92.78
Ms (P15)		45.55	0.77	32.32	4.12	0.74	0.75	10.2	94.73
Ms' (P15)		46.43	0.31	29.18	5.56	1.64	0.19	10.93	94.79
Ms' Bi (3075-2)	16	46.21	1.24	29.15	4.35	1.88	0.04	10.56	95.76
Ms″ (STL)	1	46.04	0.125	33.6	1.05	0.544	0.027	10.64	92.64
Ms″ (P. Mouton)	6	47.57	0.83	36.18	1.39	0.68	0.32	8.5	95.47
Ms″ (FC)	3	51.74	0.04	32.89	2.49	1.11	0.07	8.98	97.32

(325-6)—Huangfengling two-mica granite, Jiangxi Prov.; (3075-2)—Baimianshi two-mica granite, Jiangxi Prov.; (P15)—Old woman-Piute Mt. two-mica granite, Southeastern California, USA (Miller et al. 1981); (STL)—Shituling alkali feldspar granite, Guidong batholith, Guangdong Prov. (from Wang W G); (P. Mouton)—Port Mouton pluton, Southwest Nova Scotia (Fallon 1998); (FC)—Fucheng peraluminous granite, Jiangxi Prov. (Zhang et al. 2010)

select pure sericite to determine its chemical composition. The authors have not carried out special analysis on it. But, no common tendency might be gained due to a wide range of difference among the analytical data in Table 1.4 gathered from various researchers.

It seems that the composition of primary and replacive muscovite in different granite bodies may vary. Obviously, it is necessary to do much more microprobe analyses so as to determine whether there is a common tendency for composition disparity between the two genesis muscovite in different granites.

1.1.4 Quartzification

The term "silicification" is extensively used to describe a rock in which the silica content is increased by metasomatic alteration. The so-called silicified granite may be divided into weak-, moderate- and intensely silicified granites (Fig. 1.44). However, the so-called "silicification" is formed, in fact, by cementation of hydrothermal microquartz rather than by metasomatism. In some places, the idiomorphic growth lines in allomorphic quartz micrograins can be seen under the microscope with reduced diaphragm, indicating the filling origin of the microquartz rather than replacement (Fig. 1.45).

Actual metasomatic quartzification is obviously of hetero-oriented replacement pattern because the crystallographic lattice of the quartz is completely different from that of the feldspar being replaced.

K-feldspar can partly be replaced by quartz in Beihuan leucocratic granite, Yangjiang County, Guangdong Province. The granite rich in silica and alkali and poor in calcium, magnesium, and ferrous iron consists of albite 40 %, quartz 37 %, alkali feldspar 19 %, and protolithionite 3 %. There are a lot of quartz grains irregularly distributed inside euhedral alkali feldspar. Moreover, many inside quartz grains nearby the border have the same orientation as that of the adjacent quartz immediately outside the euhedral K-feldspar. Evidently, the inside quartz takes the orientation as that of the outside one. Therefore, the quartz (Q_1, Q_3) outside K-feldspar can be inferred as primary, while the irregular curved quartz (Q_1', Q_3') inside the K-feldspar is of metasomatic.

At the beginning of quartzification, the replacive quartz invades rapidly to the inner part of K-feldspar with its outside contour remained (Figs. 1.46 and 1.47).

Obviously, during quartzification, K-feldspar with some fine albite lamellae is easily replaced and most perthitic albite lamellae would be retained. Primary quartz, whole plagioclase, and biotite (protolithionite) survive quartz replacement.

Fig. 1.44 (+). So-called "intensely-silicified granite" is, in fact, granitic breccias cemented by microcrystal quartz. Zhuguangshan batholith, Guangdong Province

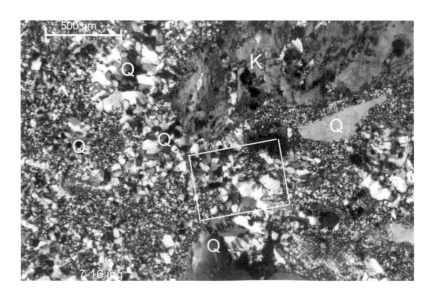

Fig. 1.45 (+)Q. Enlargement of frame in Fig. 1.44. Idiomorphic growth line in quartz reveals that quartz crystallized in free space

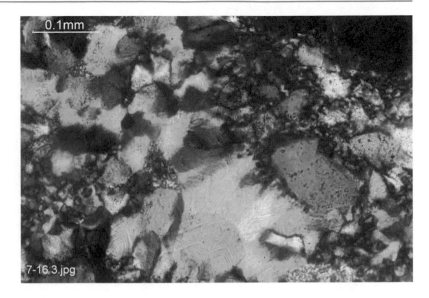

Fig. 1.46 (+)Q. Quartz $Q_1Q_2Q_3Q_4Q_5$ distributed outside the idiomorphic alkali feldspar K_1 are primary, while quartz Q_1' $Q_3'Q_4'$ irregularly growing inside the K_1 is replacive. $Q_1'Q_3'Q_4'$ have the same crystallographic orientation as that of the primary quartz $Q_1Q_3Q_4$, respectively. Beihuan leucocratic granite

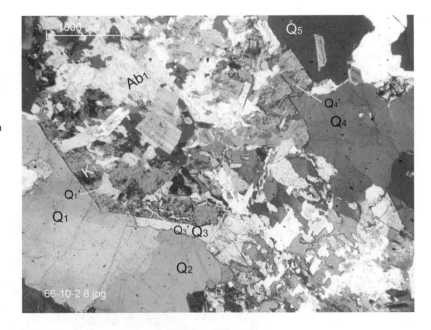

The rock with quartzification has no evident traces of deformation, showing that the quartzification occurred under a stable environment (Figs. 1.46 and 1.47). Furthermore, K-feldspar is replaced by quartz in irregular curved forms rather than along cleavages or fissures.

After intense quartzification, all K-feldspar parts are completely replaced and eliminated, the remaining minerals are skeleton relics of perthitic albite, idiomorphic sodic plagioclase (Fig. 1.48), thick flaky protolithionite, as well as, primary quartz (Fig. 1.49).

In sodium metasomatite from Jiling granite, Gansu Province the primary quartz may be replaced by calcite, which, later on, is easily replaced by the newly formed or late stage quartz (Fig. 1.50) (please refer to Sect. 3.6).

Fig. 1.47 (+)Q. Q′ replaces intensely alkali feldspar going quickly deep into the center. Beihuan leucocratic granite, Yangjiang County, Guangdong Province

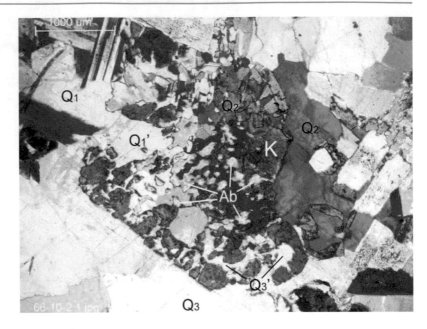

Fig. 1.48 (+)Q. Albitic plagioclase and skeleton perthitic albite survive intense quartzification. Beihuan leucocratic granite

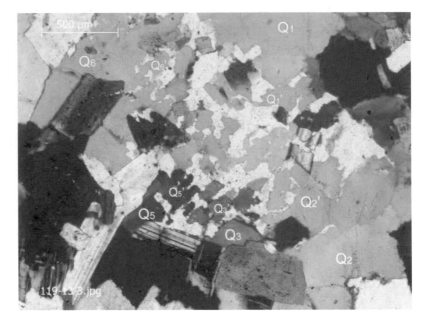

1.1.5   Protolithionitization

Protolithionite $(K_2(Fe, Mg, Mn, Ti)_{2.5}Li_1Al_2$ $[Al_2Si_6O_{20}](OH, F)_4)$ is a kind of biotite group occurring in granite rich in silica, alkaline, aluminous, and volatile component. Protolithionite contains more Li and F than common biotite and has weak pleochroism. Thick flaky protolithionite may occur in anhedral form among idiomorphic albite crystals (Fig. 1.51).

Nevertheless, some remnants of perthitic albite appear at the border of protolithionite (Fig. 1.52) or even within protolithionite (Fig. 1.53), indicating that at least a local part of protolithionite is not primary, but is formed by replacement of K-feldspar.

Fig. 1.49 (+).
Protolithionite Bi survives
quartzification. Beihuan
leucocratic granite

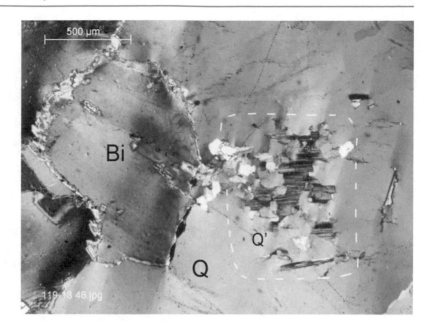

Fig. 1.50 (+). Calcite Cc
having replaced primary
quartz was later replaced by
the newly formed quartz
Q′. Sodium metasomatite in
Jiling granite, Gansu
Province

In Fig. 1.52, the protolithionites Bi′ and Bi have the same optical characteristics, i.e., the same weak pleochroism, and identical orientation. There is no clear boundary between Bi and Bi′. Unfortunately, microprobe analysis has not been done because of the cover glass on the old thin section. So, it is not clear whether there is difference in their composition.

There are two hypotheses for the genesis of protolithionite:

The first hypothesis suggests that both Bi and Bi′ have the same origin, i.e., they are metasomatic, because Bi′ replaces K-feldspar and no difference exists between Bi and Bi′.

The second hypothesis claims that Bi′ is formed by metasomatism, while Bi is probably primary.

Fig. 1.51 (+). Primary protolithionite Bi occurs interstitially among euhedral plagioclases (albite). Beihuan leucocratic granite, Yangjiang County, Guangdong Province

Fig. 1.52 A(−); B(+). Alkali feldspar is intensely replaced mainly by quartz Q′ and partly by Bi′ with perthtic albite relicts remained. Note the primary protolithionite Bi and replacive protolithionite Bi′. Beihuan leucocratic granite

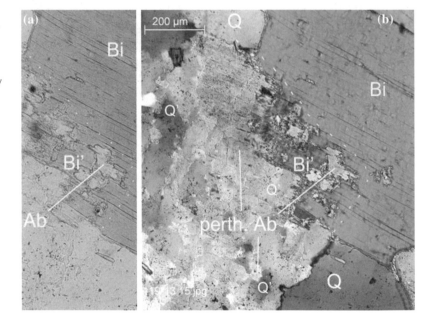

The authors agree with the second hypothesis. In other words, there are both primary and metasomatic protolithionite in the rock. Back against the primary protolithionite Bi, the metasomatic protolithionite Bi′ grows and replaces the K-feldspar. The primary protolithionite can generally be determined by thick flake (thickness is greater than 0.5 mm for the rock), lack of relics of perthitic albite, and smooth contact with quartz and/or idiomorphic plagioclase, such as Bi in Figs. 1.51 and 1.52. The metasomatic protolithionite is determined by the presence of relics of perthitic albite (Bi′ in Figs. 1.52 and 1.53), although there is no clear boundary between primary and metasomatic protolithionite.

Fig. 1.53 A(−); B(+)Q. Skeleton residues of perthitic albite in metasomatic quartz Q′ and Bi′, indicating Bi′ is of metasomatic too. Beihuan leucocratic granite

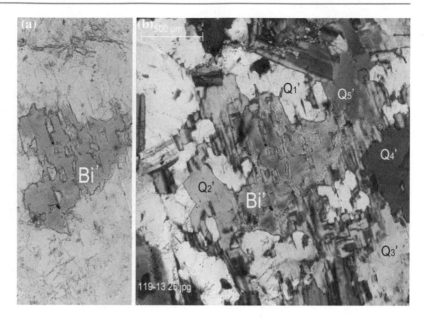

Due to limitation of thin section, the primary mica upon which the metasomatic protolithionite contacts in Fig. 1.53 may not be seen but should exist below or above the thin section.

Because perthitic albite developed pervasively in the alkali feldspar of the rock, the replacive protolithionite replacing alkali feldspar should have remnants of perthitic albite. Nevertheless, no skeletons of perthitic albite are enclosed in Bi, showing that Bi may not necessarily be metasomatic, but primary in origin.

Therefore, the authors consider that both primary and secondary or metasomatic protolithionite may exist in the rock.

1.1.6   Beryllization

Small miarolitic cavities appear in apical or periphery facies of Shanbei leucogranite, Taishan County, Guangdong Province. The leucogranite is rich in silica and alkali. In the center of such miarolitic cavities anhedral beryl (Ber) is surrounded by idiomorphic K-feldspar and albite. The anhedral beryl should be primary in origin. The beryl (Ber′) distributed irregularly inside the K-feldspar has the same crystallographic orientation as that of the anhedral beryl immediately outside the K-feldspar. So, Ber′ has a nibble

replacement origin (Figs. 1.54 and 1.55). Albite, quartz, and biotite are comparatively stable during beryllization. (Please refer to Figs. 3.149 and 3.150).

The above-mentioned hetero-oriented (nibble) metasomatic growth of rock-forming minerals occurs in granitic rocks where abundant identical or similar minerals would naturally serve as a crystallographic nucleus for nibble replacement. If metasomatism should have occurred where identical or similar minerals as a replacive mineral were absent, the impurities or defects in crystal lattice could serve as a crystal nucleus for metasomatic growth. Such examples can be indicated by the scattered growth of sericitization or even muscovitization in plagioclase, the calcitization or carbonitization, and the pyritization as follows.

1.1.7   Calcitization or Calcite Replaces Quartz and K-feldspar

In general, calcite is absent in granite, so there is no calcitization in granite. However, calcitization is observed in sodic metasomatites from Jiling granites, Gansu Province. The sodic metasomatite is characterized by loss of quartz, which is replaced first by calcite rather than by albite.

Fig. 1.54 (+)Q. Primary
beryl Ber and metasomatic
beryl Ber′. Ab and Bi
survive. Shanbei granite,
Taishan County,
Guangdong Province

Fig. 1.55 (+)Q. Replacive
Ber′ has the same
orientation as Ber outside.
Shanbei granite

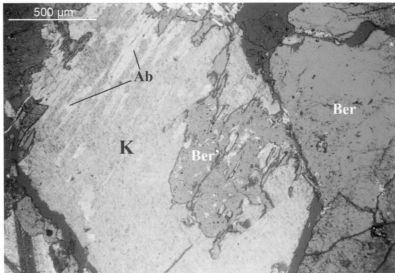

Primary quartz is initially replaced by branch-like calcite (Fig. 1.56), then becomes relicts in calcite (Fig. 1.57), and disappears finally (Figs. 1.58 and 1.59).

With further calcitization, calcite begins to replace K-feldspar in snowflake form (Fig. 1.58), and to a less extent in sericitized plagioclase, while chlorite from biotite and whole plagioclase remain unchanged.

After intense calcitization, quartz and K-feldspar are thoroughly replaced resulting in the abundance of calcite with scattered albite grains (both perthitic albite and co-oriented albitized albited from K-feldspar). The chlorite (from biotite) and pseudomorphic albite (from plagioclase) survived calcitization (Fig. 1.59).

1.1.8   Pyritization

Pyrite crystals are generally absent in fresh or unaltered granite. Perhaps a very small amount of pyrite may exist as an accessory mineral. However, many idiomorphic cubic crystals of pyrite

Fig. 1.56 (+). Calcite begins to replace Q, while K is temporarily retained. Jiling granite, Gansu Province

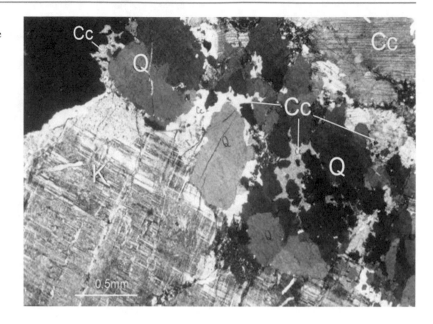

Fig. 1.57 (+). Calcite seriously replaces quartz Q, while Pl(Ab) is not replaced by calcite Cc. Sodium metasomatite in Jiling granite

($d \approx 1.5$ cm) are found in cataclastically deformed altered Jiulianshan granite, Heping County, Guangdong Province (Rong 1982). A relict of plagioclase inside the pyrite porphyroblast is in optical continuity with a plagioclase outside nearby (Fig. 1.60). Furthermore, the pyrite porphyroblast is idiomorphically formed probably due to its strong crystallization force.

1.2   Co-oriented Replacement Pattern

Co-oriented replacement means that the replacive mineral has the same or similar crystallographic lattice as that of the replaced mineral; i.e., the replacement process proceeds along the lattice

Fig. 1.58 (+). Calcite Cc irregularly replaces K in snow flake form. Jiling granite

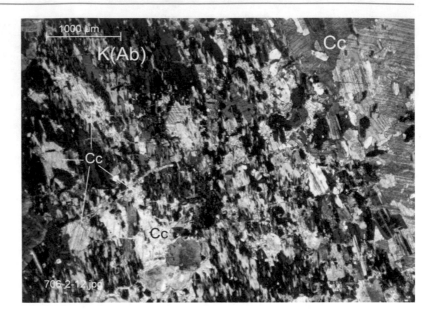

Fig. 1.59 (+)Q. Cc has replaced Q and K completely with Pl(Ab) and Bi(Chl) survived. Sodium metasomatite in Jiling granite

orientation of a replaced mineral, resulting in partial or even complete transformation of the original mineral into a new one. Included in the co-oriented replacement pattern are deanorthitization of plagioclase transformed in situ to albite, co-oriented albitization of K-feldspar, as well as muscovitization and chloritization of biotite.

1.2.1 Co-orientated Muscovitization of Biotite

Both muscovite $K_2Al_4[Si_6Al_2O_{20}](OH, F)_4$ and biotite $K(Mg, Fe^{2+})_{3-2}(Fe^{3+}, Al, Ti)_{0-2}[Si_{6-5} Al_{2-3}O_{20}]O_{0-2}(OH, F)_{4-2}$ have a common basal

Fig. 1.60 (+).
Replacement growth of
idiomorphic pyrite in
cataclastic granite. A relict
of plagioclase is seen in the
pyrite. Jiulianshan granite,
Heping County,
Guangdong Province

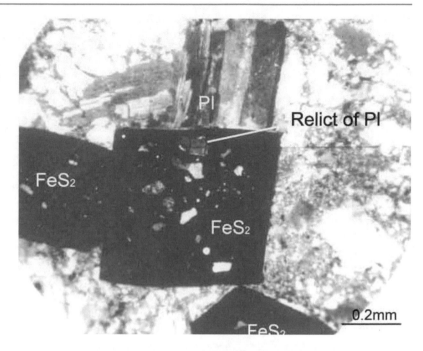

Fig. 1.61 A(−); B(+).
Biotite Bi is co-orientedly
transformed to muscovite
Ms′. Baimianshi two-mica
granite

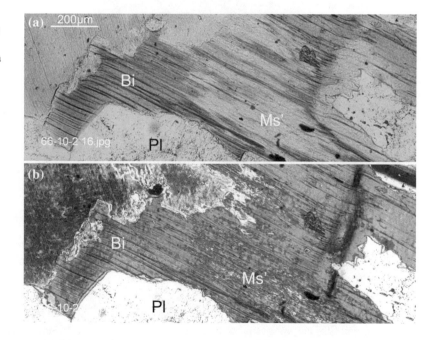

plane (001) and similar crystallographic lattices.
Hydrothermal fluid containing Si and Al
can penetrate along cleavage into biotite, displacing Fe, Mg, and Ti, resulting in transformation from biotite to muscovite (Figs. 1.61
and 1.62).

1.2.2 Co-orientated Chloritization of Biotite

The crystallographic lattice of chlorite is different
from that of biotite, but still in many respects
resembles the lattice of micas (Deer et al. 1963,

Fig. 1.62 A(−); B(+). Bi
(euhedral) co-orientedly
transformed to muscovite
Bi(Ms)(dirtier) with ferric
educt is epitaxially
enclosed by primary Ms
(clearer)

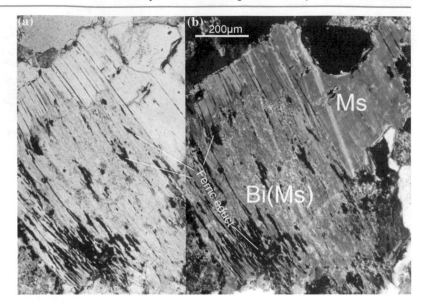

Fig. 1.63 (+). Biotite Bi is
in part transformed to
chlorite Chl. Baimianshi
two-mica granite

2001). Both contain basal cleavage (001). After
alteration, biotite is prone to be replaced by
chlorite in pseudomorphic form (Fig. 1.63).

In pseudomorphic chlorite, K_2O is thoroughly
removed, SiO_2 and TiO_2 are evidently decreased,
while MgO, FeO, and volatile component are
obviously increased as their total amounts are
decreased (Table 1.5).

1.2.3 Co-oriented Albitization or Deanorthitization of Plagioclase

Deanorthitization of plagioclase is commonly
observed in altered granite. Because of alteration
(sericitization for oligoclase and less calcic
andesine, and saussuritization for more calcic

Table 5 Microprobe analyses of biotite and pseudomorphic chlorite

Mineral	Mean	SiO$_2$	TiO$_2$	Al$_2$O$_3$	$\sum$FeO	MnO	MgO	Na$_2$O	K$_2$O	Total
Bi	3	36.4	3.49	15.68	22.1	0.43	8.23	0.082	9.52	96.1
Chl	3	28.26	0.32	16.45	27.58	0.62	12.69	0.2	0.1	86.37

Sample from Jiling granite, Gansu Province

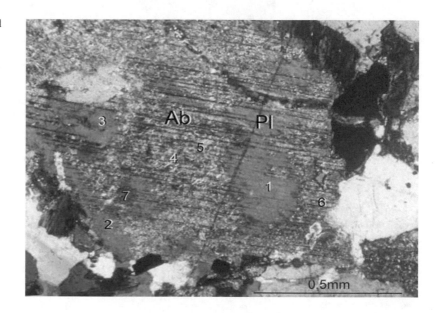

Fig. 1.64 (+). Sericitized plagioclase (4, 5, 6) is transformed to albite (An$_{2-4}$). The unsericitized plagioclase (1, 2, 3) maintains primary constituent (An$_{20-25}$) and small piece of K (7)

andesine and labradorite), the plagioclase is changed in situ into albite while preserving the original twinning.

During deanorthitization, the Na ion enters as the Ca ion is subtracted from the crystallographic lattice. Moreover, SiO$_2$ needs to be introduced. The surplus Ca and Al from the replacement process as well as other impurities move out, forming zoisite, clinozoisite, epidote, sericite (montmorillonite et al.) and even calcite in altered plagioclase. The core with higher An values of normally zoned plagioclase is generally changed to more albitic compositions while the rim composition remains unchanged. Tiny grains of topaz and fluorite contained in altered plagioclase (albite) of granite rich in F are also reported (Haapala 1997). Thus, the whole plagioclase turns into or approaches stable albite. The distribution of sericitization in plagioclase may not be homogeneous. The sericitized part of plagioclase is albitized, while the unsericitized plagioclase may retain its original composition (Fig. 1.64).

The source of potassium that creates sericite during alteration of plagioclase is inferred to come from antiperthitic K-feldspar. The authors, however, have observed that the antiperthitic blocky K-feldspar remains seemingly unchanged after comparable intense sericitization of the plagioclase (Fig. 1.65). Therefore, the source of potassium for sericitization might not have originated from antiperthitic K-feldspar, but come either from outside or from transformation of the plagioclase.

1.2.4 Co-oriented Albitization of K-feldspar (K-feldspar Is Transformed into Albite)

Alkali (mainly sodic) metasomatites, (i.e., épisyenites in European literature), occurring in peraluminous and metaluminous granitoids are characterized by disappearance or evident decrease

Fig. 1.65 (+). Plagioclase is sericitized and albitized with unchanged remnants of antiperthitic K-feldspar (with very *dark gray* interference color). Jiling granite

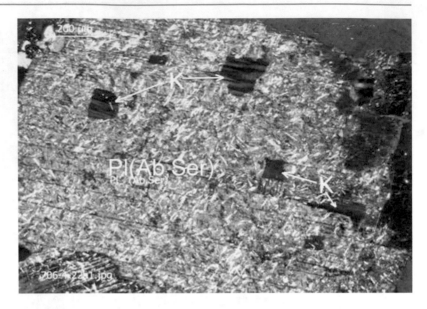

Fig. 1.66 Abrupt contact boundary (*black line*) of porphyritic granite γ with alkali metasomatite ξ (lack of quartz) which was cut across by an aplite vein. The left part of the vein in ξ also lost quartz, without evident cataclasis. Jiling granite, Gansu Province

of quartz, increase of total amount of alkali feldspars, and reservation of the basic texture. Alkali metasomatites may be subdivided into sodium (predominant), sodium–potassium (mixed), and potassium subtypes (The sodium metasomatites is only discussed in the book).

Tongue- or vein-like alkali metasomatites (tens cm to several m wide, several m to tens m long, steeply dipping) are accidentally seen in granites in southern and northern China. They are not vein or intrusion, but formed by transformation in situ from various types and multistages of granites (including pegmatite and aplite), and formed prior to the injection of latest diabase dike. The metasomatites and its surrounding granites are usually accompanied by clouded hematitization, resulting in red or brown red color. There is a gradual transition zone (1 to several meters wide) from hematitized granite to normal granite, while an abrupt contact (<1–2 cm) between alkali metasomatites and its surrounding hematitized granites (Figs. 1.66 and 1.67).

Fig. 1.67 Medium-coarse grain granite γ with gray quartz contacts abruptly (*dot line*) at alkali metasomatite ξ without quartz. The quartz was replaced by calcite, which was easily dissolved to pores under weathering condition. Jiling granite, Gansu Province

Figure 1.66 shows a complete aplite vein horizontally crossing the granite. The vein in metasomatites (left part of the figure) is also transformed into metasomatites due to lack of quartz, indicating that alkali metasomatism took place without evident fracturing after the aplite had intruded.

In normal sodium metasomatites in granites, the primary K-feldspar crystals are preserved unchanged. However, most primary K-feldspars may completely be transformed or fully co-orientedly albitized into albite in some sodium metasomatites, e.g., in Jiling granite, Gansu Province, which results postulatedly from intense sodium metasomatism, Seemingly, the intense sodium metasomatism has a special ability of transforming K-feldspar by steaming and boiling in situ into albite. The transformation from K-feldspar phase to the co-oriented albitized albite is so sudden and hidden that the transition phenomenon was not noticed in thin section observation and has been puzzled for a long time.

Nevertheless, later on, with the aid of extensive sampling and carefully observing thin sections, the co-oriented albitization of K-feldspar from weak to strong development can be found in a series of thin sections. The transition zone from unchanged K-feldspar to fully co-orientedly albitized albite ranges within 1 meter only.

The co-oriented albite from K-feldspar is characterized by the following features:
(1) Clouded hematitization.
(2) Initially dotted, then in slices, spreading to larger areas (lumps) and finally the whole K-feldspar is transformed into albite.
(3) Presence of special chessboard twins.

Co-oriented albitization of K-feldspar occurs always together with clouded hematitization (hydrogoethite). At first, the co-oriented albite takes place in form of several dots as seen in Figs. 1.68 and 1.69. The dotted co-orientedly albitized albite may be shown indistinctly in places where the clouded hematitization concentrates, implying that hematitization acts as a pioneer, followed by co-oriented albitization. Moreover, co-orientedly albitized grains first occur in the pure K-feldspar phase away from perthitic lamellae. It is even really incredible that it occurs independent of cleavage (001) as seen in Fig. 1.69. Then it takes the shape of slice along (010) in Fig. 1.70 which is different from that of perthitic lamellae mainly along ($\overline{1}502$). Furthermore, several slices nearby may combine together giving rise to dumpling and lumpy shapes (Figs. 1.71 and 1.72), which are sometimes quite special and may be distinguished from sericitized oligoclase and K-feldspar (Figs. 1.73 and 1.74). In addition, the scope of co-oriented albite is

Fig. 1.68 (+). Section ⊥ axis *b*. Co-oriented albitization (with cloudy hydrogoethite) begins with group-dotted and slice form along (010), which is different from perthitic albite. Jiling granite

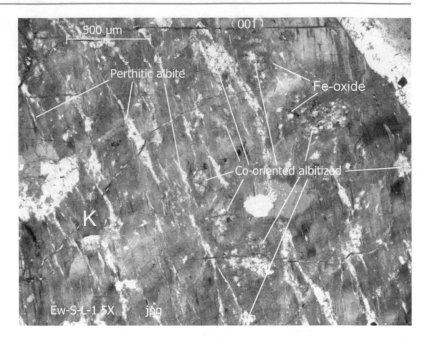

Fig. 1.69 (+). Section is ⊥ axis *b*. Co-oriented albitization (co-oriented Ab) begins to develop in K-feldspar phase, accompanied by hydrogoethite, but independent of cleavage (001). Jiling granite

distributed irregularly (Figs. 1.72 and 1.73). When it expands and includes perthitic lamellae, the whole K-feldspar crystals are completely transformed into co-oriented albite finally. A schematic diagram (Fig. 1.76) shows the different forms and distribution patterns of co-oriented albitization compared with perthitic albite lamellae (Fig. 1.75).

The boundary between K-feldspar and co-oriented albitized albite is abrupt and distinct, even in the BSE image magnified 4000 times (Fig. 1.75).

Fig. 1.70 (+). Section is roughly //axis *b*. K-feldspar is subjected to co-oriented albitization in slices or pile form, accompanied with hydrohematite (*brown* and dirtier)

Fig. 1.71 (+). Section is roughly ⊥ (010). Co-oriented albitized K (Ab) occurs in form of slice (along (010)), then cooperated to lumpy form different from perthitic albite

When thin section is ⊥ (010), the co-oriented albite commonly shows chessboard twinning (Fig. 1.77a) with variable zigzag twin compositional plane (partly being of (010)). The special chessboard twinning is distinct from either tweed (grid) twinning of microcline or well-formed parallel albite twinning of plagioclase (Fig. 1.78).

When section is //(010), the co-oriented albite reveals mottled (Fig. 1.77b) even close to uniform shapes (Fig. 1.81).

Figure 1.75 indicates that micropores as well as bright particles of hydrogoethite are contained more in the co-oriented albite than in the K-feldspar.

Fig. 1.72 (+). Section is roughly ⊥ (010). Right part of K-feldspar is co-orientedly albitized in lumpy form. Jiling granite

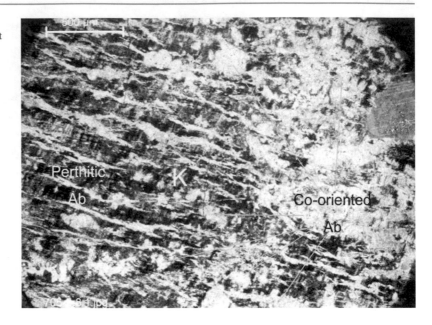

Fig. 1.73 (+). Section is roughly //axis *b*. Local co-oriented albitization occurs at contact of K-feldspar phenocryst with the enclosing oligoclase. Jiling granite

In completely co-orientedly albitized albite of K-feldspar the perthitic albitic lamellae may be observed as slightly transparent stringers (Fig. 1.78). Microprobe analyses of primary K-feldspar K and co-orientedly metasomatic chessboard albite K(Ab) are listed in Table 1.6.

Table 1.6 indicates that in co-oriented metasomatic chessboard albite, Na_2O content is greatly increased while K_2O content is nearly eliminated, and SiO_2 content must be accordingly increased. The An value of the replacive albite is less than 1.

Fig. 1.74 (+).
Enlargement of Fig. 1.73.
Co-oriented albitized albite
(*dark* and *light gray*) is
different from oligoclase
(Ser) and K-feldspar K
(*black* and *gray plaid*).
Jiling granite

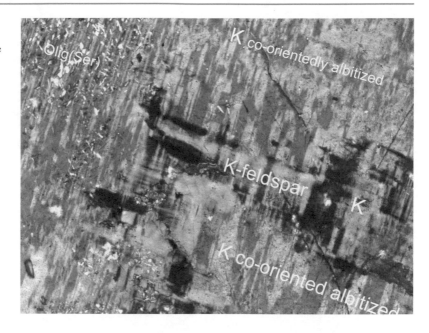

Fig. 1.75 BSE-images.
Irregular sharp boundary
between K and K(Ab).
K(Ab) contains more
micropores and hematite
Hm particles than in K.
Jiling granite

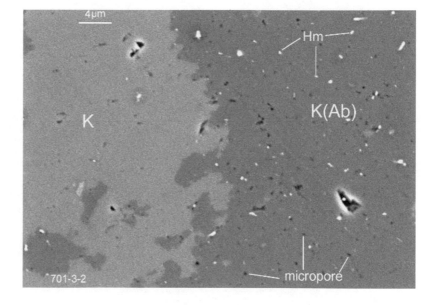

The senior author studied the optic indicatrix of two individual crystals of co-oriented chessboard albite twins by universal stage on the Wulff-net (Fig. 1.79). The result shows that the three planes connecting the homonymous axes (Np_1Np_2, Nm_1Nm_2, and Ng_1Ng_2) of optic indicatrix of two individual co-oriented chessboard albite intersect a narrow triangle column which is approximately close to axis *b* of the crystal.

Therefore, the twin axis is nearly the axis *b*, while the twin composition plane is zigzag and irregular, so the co-oriented chessboard albite twin can thus be named by the senior author as "axis-*b* twin."

After repeated examination of thin sections, the phenomena of co-oriented albitization of K-feldspar are found not only in intense metasomatites lacking quartz, but also in some

Fig. 1.76 Schematic
diagram of perthitic albite
lamellae and co-oriented
albitization. Accompanied
by cloudy hydrogoethite,
the co-oriented albitization
begins with *dotted*, slice
(along (010), then lump
forms, which are different
from the perthitic lamellae
(mainly along Murchison
plane ($\overline{1}502$) (*left figure*).
The preliminary
co-oriented albitization is
rapidly transformed to
complete one (*right figure*).
$N_pN_mN_g$—3 axes of optic
indicatrix of co-oriented
albite

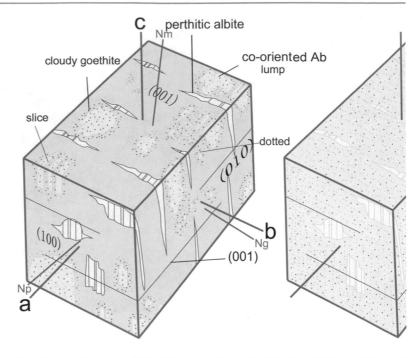

Fig. 1.77 (+). Chessboard
twin (**a**) or mottled twin
(**b**) of co-orientedly
albitized albite.
a—Section is ⊥ axis *a*;
b—Section is ⊥ axis
b. Sodium metasomatite in
Jiling granite

surrounding granites where primary quartz still
exists (Figs. 1.80 and 1.81). Later microscopic
observation indicates that the co-oriented albitization of K-feldspar is the first step of alkali
metasomatism, followed by calcitization of
quartz and K-feldspar. Later on, the calcite is
replaced by albite (see Sect. 3.6. Successive
sequence of multi-metasomatic processes in
alkali metasomatites).

1.2.5 Pseudomorphic Replacement of Chlorapatite by Hydroxy-Fluor-Apatite

Engvik (2009) studied the apatite (with thickness of
1–5 cm) in scapolitized and albitized Ødegården
metagabbro (Bamble, South Norway) that was
formerly mined as an apatite deposit (Bugge 1922).
The primary fluor-chlorapatite (nonporous and

Fig. 1.78 (−). Ghost perthitic albite in co-oriented albitized albite from K-feldspar K(Ab). A—Section is roughly // axis *b*. B—Section is nearly ⊥ axis *b*

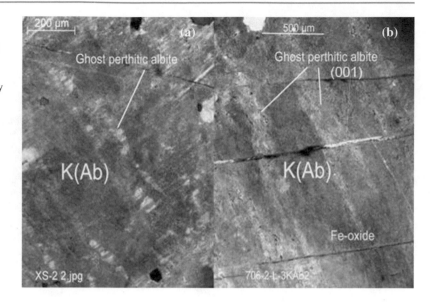

Fig. 1.79 Wulff-net stereographic projection of indicatrix axes of chessboard albite twin 1 and 2 in Fig. 1.77a. The three planes separately connecting synonym indicatrix axes of individual crystal 1 and 2 of chessboard albite twin intersect at one line, which must be the twin axis that is close to axis *b* from the stereogram

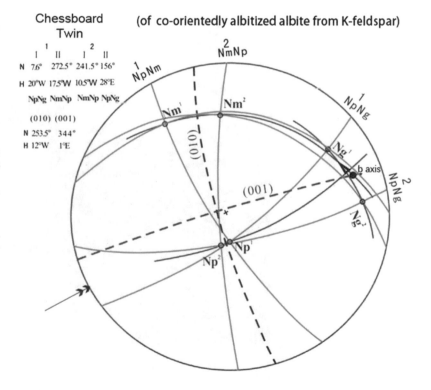

homogenous in composition with F-content as much as 3.5 wt% and Cl-content as much as 2 wt%) is transformed into chlorapatite (Cl–Ap) with Cl-content of as much as 6.8 wt% after scapolitisation. A secondary pseudomorphic replacement reaction (e.g., co-oriented replacement) which transforms chlorapatite to porous hydroxy-fluorapatite (OH–F–Ap) with only minor Cl is related to albitization (Fig. 1.82). The replacement interface is sharp on a nanoscale, observed by energy-filtered

Fig. 1.80 (+). K-feldspar is strongly co-orientedly albitized while the primary quartz is preserved

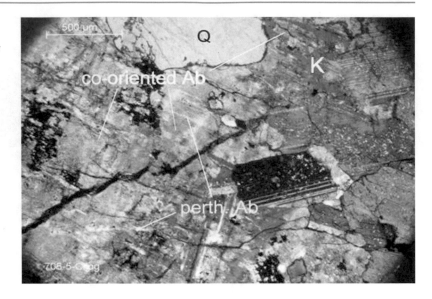

Fig. 1.81 (+). K-feldspar is thoroughly co-orientedly albitized while primary quartz survives

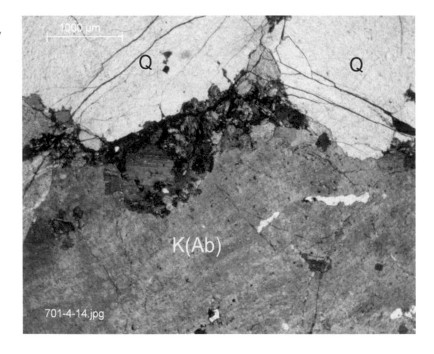

Table 6 Microprobe analyses of primary K-feldspar K and co-oriented metasomatic chessboard albite K(Ab)

Mineral	Mean	SiO$_2$	Al$_2$O$_3$	$\sum$FeO	CaO	Na$_2$O	K$_2$O	Total	An
K (701-1-73b)	5	64.27	18.39	0.07	0.01	0.65	16.04	99.9	–
K(Ab) (706-2-5a)	6	68.22	19.63	0.1	0.08	11.35	0.06	99.44	0.39

Sample from Jiling granite, Gansu Province

Fig. 1.82 BSE image. Chlorapatite (Cl–Ap) is partly co-orientedly replaced by hydroxy-fluor-apatite OH–F–Ap. From Ap+Phl vein in Ødegården metagabbro, south Norway (Engvik 2009)

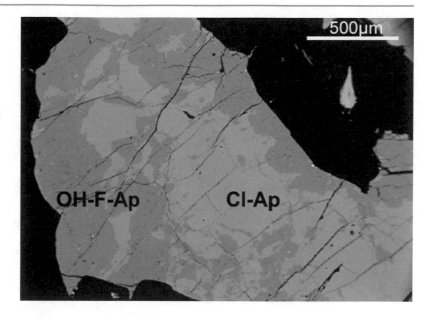

Fig. 1.83 Enlargement of BSE image. Note the higher microporosity in OH–F–Ap and trail of pores locally occurring along the replacement interface (*black arrows*) (Engvik 2009)

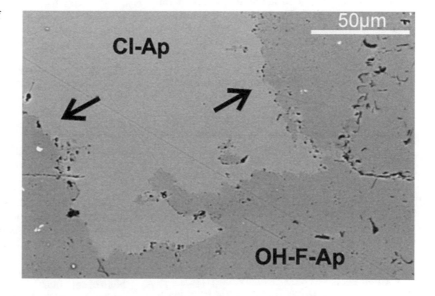

transmission electron microscopy (TEM), and the crystallographic orientation of the apatite is preserved as seen in the diffraction patterns across the interface region. There is higher microporosity in the hydroxy-fluor-apatite, and trails of pores locally occur along the replacement interface (Fig. 1.83). These observations, together with the porosity development in the hydroxy-fluor-apatite suggest that the replacement mechanism is by interface-coupled dissolution–reprecipitation (Putnis 2002).

This special metasomatic phenomenon may not be seen under optical microscopic observations because apatite has very low interference color.

Yanagisawa K et al. (1999) put grains of chlorapatite ($Ca_{10}(PO_4)_6Cl_2$) into KOH solution at 500 °C for 3 h. The result shows that the rim (40 μm wide) of chlorapatite is transformed into hydroxylapatite

$(Ca_{10}(PO_4)_6(OH)_2))$, and the boundary between rim and core is quite sharp. Abundant micropores develop in the hydroxylapatite rim. The mechanism of replacement is considered to be either by ion exchange (Yanagisawa et al. 1999) or by dissolution–recrystallization (Putnis 2002).

References

Althaus EK (1970) An experimental re-examination of the upper stability limit of muscovite plus quartz. Neus Jahrb Mineral Monatsh 325–336

Augustithis SS (1973) Atlas of the textural patterns of granites, gneisses and associated rock types. Elsevier, Amsterdam, p 447

Borodina NS, Fershtater CB (1988) Composition and nature of muscovite in granites. Int Geol Rev 30 (4):375–381

Bugge C (1922) Statens apatitdrift i rationeringstiden. Norges Geologiske Undersøkelse 110, 1–34

Carstens H (1967) Exsolution in ternary feldspars; (II) Intergranular precipitation in alkali feldspar containing calcium in solid solution. Beitr Mineral Petrog 14:316–320

Cheng YQ (1942) A hornblendic complex, including appinitic types, in the migmatite area of North Sutherland, Scotland. Proc Geol Assoc 53(2):67–85

Cheng YQ, Shen QH, Liu GH et al (1963) Some problems and working methods of metamorphic rocks. Science Press, Beijing (in Chinese)

Collins LG (1997) Sphene, myrmekite and titanium immobility and mobility; implications for large-scale K- and Na- metasomatism and the origin of magnetite concentrations. ISSN 1526-5757, electronic internet publication: http://www.csun.edu/~vcgeo005/Nr18 Lyon Mtn

Collins LG (1998) Exsolution vermicular perthite and myrmekitic mesoperthite, Myrmekite, ISSN 1526-5757, electronic Internet publication: http://www.csun.edu/~vcgeo005/Nr32Perthite[1].pdf

Collins LG, Collins BJ (2002a) Myrmekite formation at Temecula, California, revisited: A photomicrographic essay illustrating replacement textures, Myrmekite, ISSN 1526-5757, electronic Internet publication: http://www.csun.edu/~vcgeo005/Nr43Temecula.pdf

Deer WA (1935) The cairnsmore of carspharirn igneous complex. Quart. Jour. Geol. Soc. (Lond.) 91:47–76

Deer WA, Howie RA, Zussman J (1963) Rock-forming minerals.3.4. Longmans, London

Deer WA, Howie RA, Zussman J (2001) Rock-forming minerals.3.4. Longmans, London

Engvik AK (2009) Intragranular replacement of chlorapatite by hydroxy-fluor-apatite during metasomatism. Lithos 112:236–246

Fallon RP (1998) Age and thermal history of the Port Mouton pluton, southwest Nova Scotia: a combined U-Pb, $^{40}Ar/^{39}Ar$ age spectrum, and $^{40}Ar/^{39}Ar$ laser-probe study

Haapala I (1997) Magmatic and postmagmatic process in tin-mineralized granites: topaz-bearing leucogranite in Eurajoki rapakivi granite stock, Finland. J Petrol 3

Hall A (1966) A petrogenetic study the Rosses granite complex, Donegal. JP 7:202–220

He TX, Lu LZ, Li SX et al (1980) Metamorphic petrology. Geological Publishing House, Beijing (in Chinese)

Hibbard MJ (1995) Petrography to petrogenesis. Prentice Hall, New Jersey

Miller CF, Stoddard EF et al (1981) Composition of plutonic muscovite; genetic implications. Can Mineral 19:25–34

Monier G, Robert JL (1986) Evolution of the miscibility gap between muscovite and biotite solid solutions with increasing lithium content: an experimental study in the system $K_2O-Li_2O-MgO-FeO-Al_2O_3-SiO_2-H_2O-HF$ at 600 °C, 2 kbar P_{H2O} : composition with natural lithium micas. Min Mag 50:257–266

Peng CC (1970) Intergranular albite in the granite and syenites of Hong Kong. Am Mineral 55:270–282

Phemister J (1926) The geology of Strth Okykell and lower Loch Shin (Explanation of Sheet 102). Scotland Geol Sur Mem

Phillips ER (1964) Myrmekite and albite in some granites of New England Batholith, New South Wales. J Geol Soc Aust 11:49–60

Putnis A (2002) Mineral replacement reactions: from macroscopic observations to microscopic mechanisms. Mineral Mag 66:689–708

Ramberg H (1962) Intergranular precipitation of albite formed by unmixing of alkali feldspar. Neues Jahrb Mineral Abh 98:l4–34

Rogers JJW (1961) Origin of albite in granitic rocks. Am J Sci 259:186–193

Rong JS (1982) Microscopic study on the phenomenon of granite mineral replacement in petrological study(1). Geological Publishing House, Beijing, pp 96–109 (in Chinese)

Rong JS (2003) Nibble metasomatic K-feldspathization. Myrmekite, ISSN 1526-5757, electronic internet publication: http://www.csun.edu/~vcgeo005/Nr46 Rong2.pdf

Rong JS (2009) Two patterns of monomineral replacement in granites. Myrmekite, ISSN 1526-5757, electronic internet publication: http://www.csun.edu/~vcgeo005/Nr55Rong3.pdf

Saavedra J (1978) Geochemical and petrological characteristics of mineralized granites of the west centre of Spain. In: Stemprok M, Burnol L, Tischendorf G. (eds.): Metallization associated with acid magmatism. Geol Surv Czech 3:279–291

Schermerhorn LJG (1956) The granites of Trancoso (Portugal): a study of microclinization. Am J Sci 254:329–348

Schleicher H, Lippolt HJ (1981) Magmatic muscote in felsitic parts of rhyolites from southwest Germany. Contri Min Petr 78:220–224

Smith JV (1974) Feldspar minerals(2). Springer Speer, New York, p 1984

Speer JA (1984) Micas in igneous rocks. In: Bailey SW (ed.) Reviews in Mineralogy, vol.13. Micas, Mineralogical Society of America, pp 229–356

Sun T, Chen PR, Zhou XM et al (2002) Strongly peraluminous granites in eastern Nanling Mountains, China: Study on muscovites. Geol Rev 48(5):518–525 (in Chinese)

Tuttle OF (1952) Origin of the contrasting mineralogy of extrusive and plutonic salic rocks. J Geol 60:107–124

Tuttle OF, Bowen NI (1958) Origin of Granite in the Light of Experimental Studies in the System $NaAlSi_3O_8$-$KalSi_3O_8$-SiO_2-H_2O. Geol Soc Am Mem 74:153

Voll G (1960) New work on petrofabrics. Liverpool Manchester Geol J 2:503–567

Wang RM (1989) Metamorphic petrology. Geological Publishing House, Beijing (in Chinese)

Wang X, Yao XJ, Wang CS (2006) Characteristic mineralogy of the Zhutishi granite: implication for petrogenesis of the late intrusive granite. Sci China Ser D-Earth Sci 49:573–583

Webster JD, Duffield WA (1991) Volatiles and lithophile elements in Taylor Creek Rhyolite: constraints from glass inclusion analysis. Am Min. 76:1628–1645

Yanagisawa K, Rendon-Angeles JC, Ishizawa N et al (1999) Topotaxial replacement of chlorapatite by hydroxyapatite during hydrothermal ion exchange. Am Mineral 84:1861–1869

Yorder HS, Eugster HP (1955) Synthetic and natural muscovites. Geochim Et Cosmochim Acta 8:225

You ZD, Zhong ZQ, Tang ZD et al (1996) Corrosion-reaction margin with inversion of polysynthetic twinning of plagioclase in migmatites. Earth Sci J China Univ Geosci 21(5):513–518 (in Chinese)

Zhang BT, Wu JQ, Lin HF et al (2010) Petrological discrimination between primary and secondary muscovites and its geological implications: a case study of Fucheng peraluminous granite pluton in southern Jiangxi. Acta Petrologica et Mineralogica 29(3):225–234 (in Chinese)

Formation Mechanisms for Mineral Replacement

2

Formation mechanisms for mineral replacement are dissolution–precipitation (crystallization) and ion exchange.

Co-oriented replacement of a certain mineral may arise from appropriate hydrothermal fluids entering along grain boundary with tiny cracks. For hetero-oriented replacement, however, two additional prerequisites are required: (1) a mineral prone to be replaced on one side; (2) a same or similar mineral as replacive one on the other side, acting as nucleation center for its growth. Without the second prerequisite, the hetero-oriented replacement would hardly occur.

The passage for hydrothermal fluids from outside is discussed first.

2.1 Passage for Gas and Liquid from Outside

For entering and moving of hydrothermal fluids from outside, the largest openings must be fracture fault zones. The smaller openings are joints and fissures in rocks, while the tiniest openings are grain boundaries, cleavages and micropores in minerals.

Many researchers consider that the rock should be broken to provide a space for hydrothermal fluid to enter and then to cause metasomatism. Some researchers (e.g., Collins L.) believe that the rocks had once been broken, but then recovered, i.e., the traces of deformation were fully eliminated after recrystallization.

This idea is quite doubtful. The texture of the rock subjected to metasomatism, however, is quite complete rather than cataclastic. Mineral replacement develops widely in entirely solid rocks, not only in fracture zone or along joint and its sides. The granites that have been subjected to multiple metasomatism might not have been broken before.

General speaking, under compression stress, quartz is easily affected, resulting in undulatory extinction from weak to strong, and biotite is deformed, while feldspars are more stable. Under stronger stress, smaller crystals may be crushed to aggregates, and cleavage of biotite and twinned lamella of plagioclase may be curved or disrupted, while K-feldspar, especially larger grains, remain nearly unchanged because of its higher compressive strength and tenacity, and only peripheral corners may be abraded and ground. Nevertheless, whole granitic rocks are still characterized by their original magmatic hypidiomorphic granular textures although having been subjected to tectonic and metasomatic processes

© Science Press and Springer Science+Business Media Singapore 2016
J. Rong and F. Wang, *Metasomatic Textures in Granites*, Springer Mineralogy,
DOI 10.1007/978-981-10-0666-1_2

Furthermore, no triple junction textures[1] typical for recrystallization occur throughout the whole rock. Therefore, the absence of hints of cracks cannot be simply explained by their elimination due to recrystallization. At most there is coalescence phenomenon in weakly broken quartz.

Microscopic observation indicates that the real replacement phenomena have surely occurred at one side or both sides of grain boundaries without obvious traces of deformation. So we can not but admit that grain boundaries and cleavages, as well as micropores are the passages available for a gas–liquid to permeate and circulate.

However, the width of tiny microcrack along grain boundary is too narrow to be measured by using an ordinary optical microscope. It has been studied by various researchers using transmission electron microscope (TEM) and high-resolution electron microscope (HREM). The width of a measured grain boundary ranges from <100 nm (Behrmann 1985), 3–5 nm (Farver and Yund 1995) to 0.5 nm (Hiraga et al. 1999).

It is hard to imagine that hot gas–liquid fluid from outside can still enter such a narrow grain boundary (only at nanometer scales), giving rise to metasomatism. However, if you pay attention to the polished granite slate stone paved on floor in a building, you may occasionally or even frequently find the dark trace of water- or oil-soaked stain dispersed irregularly at the marginal part of the slate (Fig. 2.1). Although the depth of stain is unknown, it surely means at least that oil or water may infiltrate into such compact granite slate stone even nowadays, much more the hydrothermal fluid under high pressure and temperature. The time needed for the gas–liquid fluid to extend through solid rock and to cause metasomatism must be quite long. However, this time interval makes up just a short episode in the long history of geological evolution. Perhaps, simulation experiment for

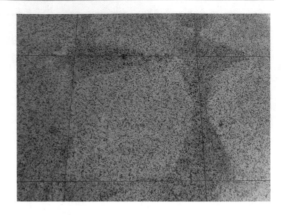

Fig. 2.1 Dark trace of oil stain appears on the dry grand slabstone of granite, showing that dirty oil-water may soak (infiltrate) in such compact slabstone nowadays, though the depth is unknown

metasomatism in compact granitic rocks may be carried out in the future.

Why does the hetero-oriented metasomatic process occur only at the grain boundaries between minerals of different kinds with different orientations rather than that between minerals of the same or similar crystallographic lattice and with the same orientation?

Grain boundaries serving as openings for entering and circulation of hot gas and fluid can be understood by the following explanation.

Figure 2.2 shows that the grain boundary (contact) between two minerals A and C is coherent (tight) when they are of the same or similar kind with the same crystallographic

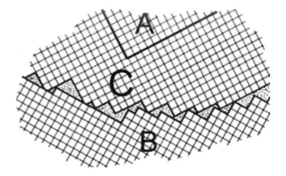

Fig. 2.2 The grain boundary between A and C is tight and coherent as they have the same crystallographic orientation, while a boundary crevice exists between B and C either they have different orientations or they are different minerals

---

[1]Triple junction texture means that adjacent rock-forming minerals (feldspars and quartz) contact each other by straight planes with the apical angle of 120° in thin section. The texture is typical for recrystallization of high degree metamorphic rocks.

orientation, while the grain boundary between minerals C and B must have tiny openings because they are of different kinds or they are situated with different orientations even if they are of the same or similar kind. Therefore, we postulate that there must be subtle crack at the boundary between minerals c and b as a passage for hot gas–fluid to enter rather than at the boundary between minerals A and C.

2.2 Mechanism of Dissolution–Reprecipitation

Dissolution–precipitation is the most general mechanism which is appropriate for explaining hetero-oriented (nibble) replacement pattern of either different kinds or the same (or similar) kind of minerals.

The dissolution–precipitation mechanism implies that a new mineral precipitates (crystallizes) in the space where an old mineral was locally dissolved during metasomatism.

It means that in the process of metasomatism an old mineral dissolves partially and gradually, while a new mineral precipitates immediately and simultaneously at the right place. The volume of the precipitated mineral is equal to that of the dissolved mineral.

For dissolution–precipitation mechanism some researchers suggest that as a replacive (guest) mineral grows, it must exert an excess pressure (Ostapenko 1976), an induced stress (Carmichael 1986), a force of crystallization (Maliva and Siever 1988), or a growth-driven stress (Merino et al. 1993) on its adjacent (host) mineral to promote its dissolution.

Merino et al. (1993) stated that in a rigid rock, the growth-driven stress between the guest and host minerals has two immediate effects: (1) an increase in the solubility and rate of pressure solution of the host mineral and (2) a decrease in the rate of growth of the guest mineral (due to decrease of the supersaturation because of increase in the solubility of the guest mineral) so that the value of the two volumetric growth-rates of guest B and dissolution of host A become equal.

Merino et al. seemed to stress that appropriate dissolution of a replaced mineral is promoted by a force driven by growth of the replacive mineral. They even emphasized: "any mineral may in principle replace any neighbor, provided there is sufficient affinity for its growth, regardless of whether the two have an element in common, of whether the two are isostructural or not, and of which mineral has the greater formula volume" (Merino and Dewers 1998).

2.2.1 Different Degree of Hetero-Orientation Replacement for Various Minerals

According to the senior author's observation, during hetero-oriented replacement, some minerals are easily replaced and some are not, while others are even barely replaced (Table 2.1), depending on their different stability and solubilitiy under a certain metasomatic condition.

Most common metasomatic processes in granite are albitization, K-feldspathization, quartzification, and muscovitization (including sericitization in plagioclase).

Table 2.1 Several replacive minerals and easily to hardly replaced minerals in hetero-oriented metasomatism

Replacive minerals	Easily replaced	Not easily replaced	Hardly replaced
Ab Q	K Cc	Perth.Ab	Pl Q Bi
Ber	K	Perth.Ab	Pl Q Bi
K	K	Perth.Ab Pl	Pl Q Bi Hb Cpx
Ms	K Bi	Pl	Q
Ser	Pl	K	Q
Cc	Q K	Perth.Ab	Pl Bi

Ab Albite; *Q* Quartz; *Ber* Beryl; *K* K-feldspar; *Ms* Muscovite; *Ser* Sericite; *Cc* Calcite; *Pl* Plagioclase; *Perth.Ab* Perthitic albite; *Hb* Hornblende; *Cpx* Clinopyroxene

K-feldspar is always by far the most easily replaced mineral in nearly all metasomatism, such as albitization, K-feldspathization, quartzification, muscovitization, and beryllization. In some places, a few tiny perthitic albites in K-feldspar may also be replaced. Nevertheless, quartz, whole (most or all) plagioclase (albite) and biotite (chlorite), as well as the majority of perthitic albite remain unchanged.

Many researchers concluded that the solubility of calcite is inversely proportional to that of quartz according to different pH values. Figure 2.3 shows that quartz and calcite have opposite solubility in different pH solutions (no matter whether fresh water or sea water). In alkalic solution, the solubility of quartz is higher while that of calcite is lower and vise versa. Therefore, quartz is replaced by calcite in alkaline condition, and calcite may be replaced by quartz when the hydrothermal condition becomes acidic.

In earlier evolution of alkali metasomatite, primary quartz is replaced by calcite from partly (Figs. 1.56 and 1.57) to completely (Fig. 1.59) during alkaline circumstances. Later on, when hydrothermal solution becomes acidic, calcite may be replaced by newly formed quartz (Fig. 1.50) as

indicated by contrary change of solubility in quartz and calcite along with the variation of pH (Fig. 2.3). It shows that the more alkalic the solution, the higher solubility of quartz and the lower solubility of calcite. On the contrary, the more acidic the solution, the higher solubility of calcite and the lower solubility of quartz.

In general case of hetero-oriented metasomatism, K-feldspar is an easily replaced mineral, indicating that it has higher solubility than other minerals. Later on, however, an inverse fact appears that plagioclase is replaced by K-feldspar, showing that the solubility of plagioclase becomes higher than that of K-feldspar in a new circumstance.

What circumstances or conditions effectively force the inverse change of solubility of the above two minerals, i.e., from a replaced mineral to a replacive mineral? That is the puzzling question to be investigated and solved.

A further mysterious fact is that old K-feldspar may hetero-orientedly be replaced by new K-feldspar, which has only slightly higher Na_2O and less K_2O content than that of the old one (see Table 1.3). Surely it does not mean that the solubility of the old K-feldspar is higher than

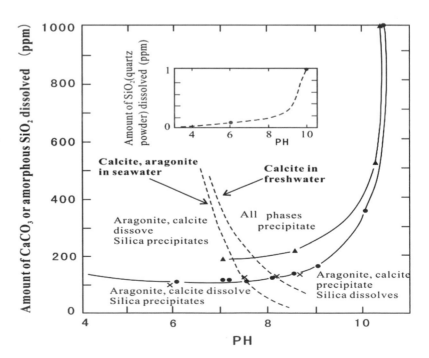

Fig. 2.3 Effect of pH at approximately 25 °C on the solubility of $CaCO_3$, quartz, and amorphous silica (Friedman et al. 1974). *Solid lines* stand for amorphous silica; other lines are labeled. (Data points from Alexander et al. 1954, *dots*; Krauskopf 1956, *crosses*; Okamoto et al. 1957, *triangles*. *Dashed lines* for aragonite and calcite from Correns 1950; for quartz from Heydemann 1966)

that of the new one? Is there any other condition affecting the solubility? What is the actual reason? These questions are not clear yet.

What is the real physical–chemical circumstance or condition controlling the change of solubility of the two kinds of feldspars and even of the same K-feldspars, whether it is old and new? This is an enigma for petrologists and mineralogists.

2.2.2 Presence of Crystal Nucleus on Which the Replacive Mineral Can Epitaxially Grow

The second prerequisite is the presence of crystal nucleus on which the replacive mineral can epitaxially grow. Since replacive minerals basically are the major rock-forming minerals, such as plagioclase, K-feldspar, quartz, muscovite, protolithionite, which are extensively or locally distributed in granites, the second prerequisite can easily be reached.

However, carbonate mineral like calcite is absent in granite. Why can the calcitization really happen? How can it be explained reasonably?

As a conjecture, an impurity probably plays a role as nucleus center. Upon the impurity a crystallite of replacive mineral can be produced. Once the initial crystallite is formed, upon which the replacive mineral would keep on growing successfully. In other words, only when a metasomatic process should have occurred, while its same or similar mineral is absent, an impurity (interstitial foreign element, etc.) may act as a nucleus center instead, such as calcitization in alkali metasomatic granite, pyritization in cataclastic granite and sericitization in plagioclase.

Generally speaking, however, as a rock-forming mineral, the replacive mineral would only take the same or similar one as its nucleus center rather than impurity. Perhaps, it is because the same or similar mineral that acts as nucleus basis much easier than impurity.

2.2.3 Explanation of Nibble Replacement of Albitization

According to the rule of dissolution–precipitation (nibble replacement), albitization can be explained as follows:

When Na-rich gas–liquid enters the boundary between K-feldspar and differently oriented plagioclase, the nibble replacement of albitization is initiated by albite growth. Since plagioclase is stable and K-feldspar is unstable and soluble, the replacive albite cannot replace the plagioclase, but unilaterally replaces the K-feldspar, forming a so-called "clear rim" of albite, which is taking the same crystallographic orientation as the adjacent plagioclase.

As Na-rich gas–liquid penetrates the boundary between two K-feldspars K_1 and K_2 (differently oriented), the replacive albite can grow at either K_1 or K_2. The direction of replacement depends on either the stronger of the crystallization ability of replacive albite or the less energy needed to dissolve the K-feldspar lattice. Both alternatives are possible and it is hard to be judged. However, the width of the nibbly replacive albite can be measured. According to the senior author's extensive observations, the width of replacive albite along axes a and c is always greater than that along axis b, resulting in an irregular flat-plate form which is much similar to the form produced by the crystallizing growth habit of plagioclase in magma, i.e., the relative growth velocities along the axes are $a \geq c \gg b$. Of course, the outline of a replacive mineral growing in the solid state is certainly irregular.

These observations precisely confirm the points of view of Merino (1988), Maliva and Siever (1988), and (Carmichael 1986) that a growth-driven stress of replacive mineral on its neighbor (host) mineral may promote its dissolution. Accordingly, can it be deduced that the growth dimension of replacive mineral is proportionally related to its growth-driven stress and is seemingly independent of the energy needed to dissolve the replaced mineral?

The surface of the grain boundary into which a gas–liquid infiltrates can be named as a "metasomatic active front" or "reaction interface" (Rong 1982, 2009). As the newly formed replacive albite crystallographically coincides with the K-feldspar on which the albite epitaxially grows, once a metasomatic albite is formed along crack, the original crack seals and disappears, and the nibble replacement can no longer take place there. The crack is transferred forwardly. Because the crystallographic orientations of the replaced K-feldspar and the nibbly replacive albite are invariably discordant, there is always an opening between them. Therefore, the metasomatic active front would continuously advance toward the replaced K-feldspar during the process. The neighboring metasomatic active fronts can be merged together into one row and march forwardly. Thus, so-called "swapped rims (or rows)" of albite can be formed (Fig. 2.4).

The fact that a clear albite rim is present along the periphery of some small plagioclase inclusions in a K-feldspar megacryst shows that Na-bearing fluid can surely penetrate into K-feldspar along its cleavage. However, the fact that no replacive albite occurs along the cleavage

results probably from lack of much more energy needed to dissolve the K-feldspar along both sides of the cleavage with neat and order arrangement of crystal lattice. Besides, it does not conform to hetero-oriented replacement regulation.

No nibble replacement growth of albite occurs at boundaries between K-feldspar and quartz. This is because the replacive albite cannot grow on quartz, although K-feldspar can be replaced by albite, meanwhile quartz cannot be replaced by albite, although the replacive albite can grow on K-feldspar. With the same reason, nibble replacement of albite cannot occur at boundaries between two plagioclases, plagioclase and quartz, plagioclase and biotite, or between other minerals.

There are sufficient minerals that may act as substrate for nibble replacement in rocks. So, it is unnecessary and unlikely for a replacive mineral to select an impurity as a nuclear center. If there are no identical or similar minerals in rocks, or the identical or similar minerals are absent at an adjacent side, then an impurity or a crystal lattice defect would act as a substrate, e.g., the sericitization or muscovitization in plagioclase and the calcitization in granite.

(a)   (b)

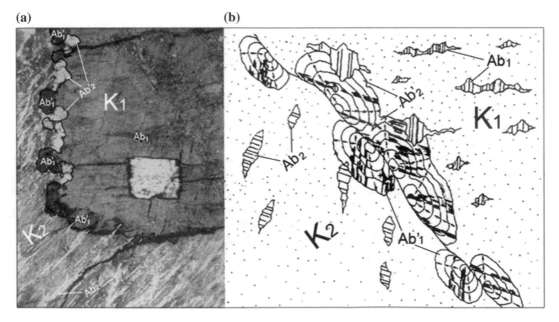

Fig. 2.4 Growth of swapped rows of replacive albite. **a** (+). Preliminary growth of swapped rows of albite. **b** Diagram of successive growth of swapped rows of albite

2.3 Mechanism of Ion Exchange or Substitution

Ion exchange mechanism means that the old mineral is partly or even wholly transformed into newly replacive mineral by the ion exchange without partial or whole dissolution of the old mineral.

The ion exchange mechanism is appropriate for the co-oriented replacement patterns of sheet minerals, such as the muscovitization and chloritization of biotite.

The structures of biotite and muscovite are similar. They have a common basal plane cleavage (001) and similar crystallographic lattices. Hot gas–liquid fluid containing Al^{3+}, Si^{4+} can penetrate along cleavage into biotite, substituting Fe^{2+}, Mg^{2+}, and Ti^{4+}, causing a pseudomorphic transformation from biotite to muscovite (Figs. 1.61 and 1.62).

The crystallographic lattice of chlorite is different from that in biotite, but still resembles the lattice of mica in many respects (Deer et al. 1963). After alteration, biotite is prone to be replaced by chlorite in pseudomorphic form (Fig. 1.63).

In recent three decades, the transition processes from biotite to chlorite have been studied by many researchers (Veblen and Ferry 1983; Banes and Amourc 1984; Kogure and Banfield 2000) (Fig. 2.5). According to observation and analysis during mapping by use of high-resolution (atomic) transmission electron microscopy, two kinds of transition mechanisms from biotite to chlorite are found: a. Mechanism 1. Growth of a brucite-like $(Mg(OH)_2)$ layer into the interlayer (K) between two TOT mica layers. Mechanism 1 requires introduction of substantial material and results in an increase in volume. b. Mechanism 2. Formation of a brucite-like layer by removal of the two tetrahedral sheets $((Si, Al)O_4)$ from one TOT mica layer. Mechanism 2 requires a net removal of material from the crystal and results in a decrease in volume. Veblen and Ferry (1983) estimated that approximately 85 % of chloritization is produced via mechanism 2 and 15 % via mechanism 1. Therefore, the final volume of chlorite basically remains unchanged.

On the basis of microprobe analysis on pseudomorphic chlorite from biotite sampled

(a)

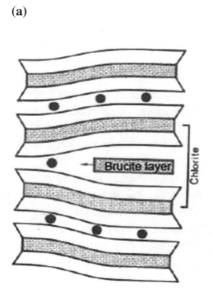
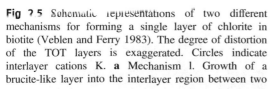

(b)

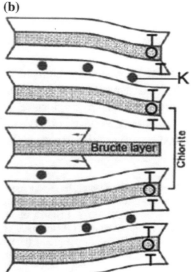

Fig 2.5 Schematic representations of two different mechanisms for forming a single layer of chlorite in biotite (Veblen and Ferry 1983). The degree of distortion of the TOT layers is exaggerated. Circles indicate interlayer cations K. **a** Mechanism 1. Growth of a brucite-like layer into the interlayer region between two TOT mica layers results in an increase in volume. Direction of growth of the layer is indicated by the arrow. **b** Mechanism 2. Formation of a brucite-like layer by removal of the tetrahedral sheets of one TOT mica layer results in a decrease in volume. Direction of dissolution of the tetrahedral sheets is indicated by the *arrows*

from Jiling granite, Gansu Province (Table 1.5), MgO and $\sum$FeO contents in chlorite have increased to 27.58 and 12.69 %, compared with 22.1 and 8.23 %, respectively, in the original biotite, which indicates that not only brucite-like layer $Mg(OH)_2$ but also $Fe(OH)_2$ had participated in substitution for the interlayer K and the two tetrahedral sheets $((Si, Al)O_4)$ during chloritization.

The replacive mineral formed by ion substitution naturally maintains the original orientation of the host replaced mineral.

2.4 Co-oriented Replacement of Feldspar Minerals

With reference to the albitization of plagioclase, it generally means that plagioclase is replaced by albite co-orientedly rather than hetero-orientedly. On the contrary, the albitization of K-feldspar normally implies that K-feldspar is replaced by albite hetero-orientedly (nibbly) rather than co-orientedly.

However, plagioclase, especially small grains or lamellae, may be hetero-orientedly replaced partly by albite. K-feldspar may also be co-orientedly replaced by albite in sodium metasomatites.

All the co-oriented replacement phenomena of feldspar minerals, in the authors' opinion, belong only to albitization from either plagioclase or K-feldspar. The authors have never observed a real example of co-oriented replacement of plagioclase by K-feldspar.[2]

---

[2]Co-oriented K-feldspathization of plagioclase was mentioned by several researchers (Collins 1998, 2003; Collins and Collins 2002; Putnis et al. 2007). Labotka et al. (2004) carried out an experiment study of replacement of albite by K-feldspar. Amelia albite powder grains (about 50–200 μm in diameter) were used to react with ^{18}O-enriched KCl (aq) (1 and 2 mol) at 600 °C and 200 MPa. After 6 days, the 150 μm diameter albite grains had 5–20 μm rims of K-feldspar in which the O was strongly enriched in ^{18}O. The contact between the core albite and the rim K-feldspar is sharp and decorated with numerous pores of various sizes. Although it is considered as pseudomorphic replacement, the reaction rim of K-feldspar (reaction product) in respect to the optical continuity with the host (Amelia albite) is still unknown.

2.4.1 Co-oriented Albitization of Plagioclase

Plagioclase is easily altered by hydrothermal fluid. After alteration, oligoclase and/or sodic andesine become blurred and sericitized, while more calcic andesine and/or labradorite become saussuritized, and the whole plagioclase is co-orientedly transformed into albite.

2.4.2 Co-oriented Albitization of K-Feldspar

Generally, K-feldspar is barely altered by hydrothermal fluid. The co-oriented albitization of K-feldspar may occur only during intense sodium metasomatism and may not appear in ordinary alkali metasomatite.

2.5 Presence of Micropores in Feldspar Is Probably One of the Key Factors to Allow the Co-oriented Replacement

In the last 30 years more detailed studies of feldspars (mainly plagioclase) in common igneous rocks have been carried out by using electron microscope, scanning electron microscopy, and high-resolution transmission electron microscope. These studies have been made either on fractured surfaces (Dengler 1976; Que and Allen 1996; Wang and Liu 2009) or on polished and ion-milled surfaces (Montgomery and Brace 1975) and have shown the presence, in addition to micro cracks, of numerous now-empty micropores[3] (within altered plagioclase) (Fig. 2.6a).

Small crystals of sericite have grown in the micropores, and the plagioclase surrounding the pores is transformed into albite. The clear part of plagioclase, mainly in the rim, but also in the

---

[3]There are different opinions on the origin of micropores: original (Roedder and Coombs 1967; Montgomery and Brace 1975); secondary (Smith and Brown 1968; Parsons 1978); both original and secondary (Que and Allen 1996).

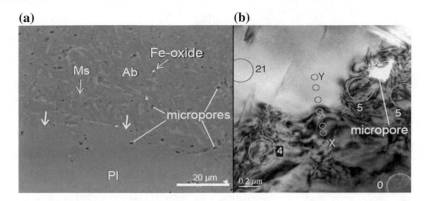

Fig. 2.6 BSE image and bright-field TEM image illustrating oligoclase–albite transformation (Engvik et al. 2008). **a** BSE image showing the replacement of oligoclase Pl (An22) by albite Ab (An2). The replacement interface (*white arrows*) is sharp. Note that oligoclase appears texturally homogeneous without pores and that the albite part of the crystal has a high microporosity. *Light*, very fine-grained laths in the albite are white mica Ms. **b** Bright-field TEM image of an untwined region featuring a replacement interface between oligoclase (*upper light gray*) and albite (*lower dark gray*). A set of analyses was made along the line between X and Y

inner part, free or less amount of either micropores or sericite crystals, retains its original composition. Perhaps, the presence of micropores is a key factor that allows the alteration of plagioclase.

Micropores ranging from several nm to μm in size (Worden et al. 1990) are irregular in shape and commonly elongate, slot-like (Fig. 2.7) which are generally considered to be interconnected. The width of transition from unaltered plagioclase to altered albite is only 40 nm (Fig. 2.6b) (Engvik et al. 2008; Hövelmann et al. 2010).

Micropores preferentially developed in the core of plagioclase with the clear margin free of micropores.

Worden et al. (1990) considered that micropores are abundant in coarsened areas of perthitic lamellae, in which porosities may be as much as 4.5 %.

Lee et al. (1995) reported that in alkali feldspar, micropores are distributed by pairs along the borders of albite lamellae (Fig. 2.8).

According to the authors' observation, micropores in K-feldspar basically occur in places of intense argillation alteration where dense fine perthitic albite lamellae have formed

(Figs. 2.9 and 2.10). Micropores are independent of cleavage (001).

What is the mechanism for feldspar to be co-orientedly replaced by albite?

The sharp and abrupt contacts between unaltered plagioclase and co-orientedly replaced albite seem hardly to be explained by the ion exchange mechanism. If co-oriented albitization is explained by dissolution–precipitation mechanism, it would be in contradiction with the rule of hetero-orientation replacement, because the latter should not take place if the crystallographic orientations of both replacive and replaced minerals are in accordance with each other. The authors are very much puzzled by this contradiction. The actual mechanism awaits further investigation.

Putnis et al. (2007) reported that there are many micropores several hundred nm in diameters in the newly formed K-feldspars which have replaced plagioclase (Fig. 2.11). Micro-hematite crystals occur in the micropores resulting in reddening of the K-feldspar.

In K-feldspar, however, there is nearly no sericitization, although micropores are also present in it.

As for the genesis of micropores in feldspar, there is primary origin (Roedder and Coombs

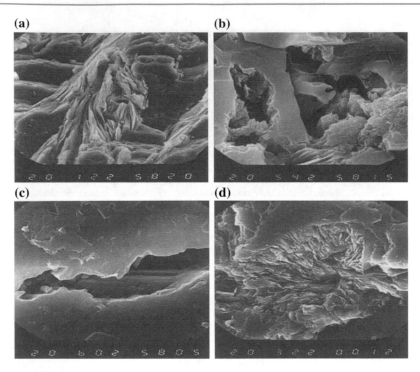

Fig. 2.7 Scanning electron micrograph (secondary electron image) of unpolished plagioclase fragment (Que and Allen 1996). **a** Sericite, altered plagioclase and micropores in a plagioclase grain, ×1000 (Fig's width 65 μm). **b** Micropores in plagioclase, ×4500 (Fig's width 14 μm). **c** Elongated clot-like micropores in plagioclase. Smaller pores at depth within the main pore also occur, ×5000 (Fig's width 13 μm). **d** Sericite flakes in a pore of a plagioclase grain

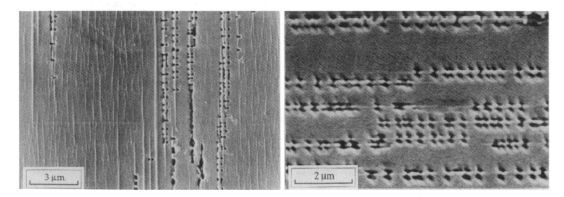

Fig. 2.8 SE image of a HF-etched K-feldspar fragment. *Left* image is of (001) cleavage fragment. Lamellae greater than ∼100 nm in width is decorated by pairs of etch pits. Narrower lamellae have few or no etch pits. The platelet-rich area (*left*) and some of lamellae are completely free of etch pits. *Right* image is of (010) cleavage fragment. Pairs of nm-sized etch pits with a distinctive twofold symmetry straddle the albite lamellae (Lee et al. 1995)

1967; Montgomery and Brace 1975), as well as secondary origin.

There are two hypotheses for the secondary origin of micropores: (1) the secondary micropores are caused by release of coherent elastic strain energy between K-rich and Na-rich phases in semicoherent microperthite. So, micropores are present at both sides of perthitic

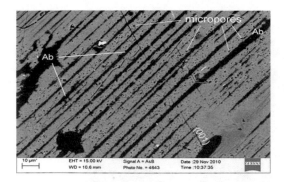

Fig. 2.9 BSE image of K-feldspar. Micropores in K-feldspar appear mainly at the adjacent area around the microperthitic albite lamellae and absent or much less present at the area of nonperthite and cryptoperthite (*lower right* and *upper left*). There is no relation with the cleavage (001). Huangnitian quartz syenite

lamellae and are absent at the periphery of coherent cryptoperthite (Brown and Parsons 1993); (2) hydrothermal fluid (bearing $NaSiO_3^-$) entered into feldspar crystal along microfissures, causing replacement. The micropores are formed because the molar volume of the newly crystallized feldspar is not equal to, but slightly less than that of the dissolved old feldspar (Worden et al. 1990; Putnis 2009; Putnis et al. 2007; Hövelmann and Putnis et al. 2010). The microfissures disappeared by coalescence later

on and left fluid inclusions are distributed linearly.

The mechanism for co-oriented replacement of feldspar has not been identified so far. There are still two hypotheses: (1) ion exchange mechanism; (2) dissolution–precipitation mechanism.

According to the hypothesis of ion exchange, the co-oriented albitization of both plagioclase and K-feldspar is formed by substitution of metallic ions within the unchanged aluminosilicate crystalline lattice.

The co-oriented albitization of plagioclase is produced by substitution of Ca^{2+} by Na^+ with slight addition of Si^{4+}. The co-oriented albitization of K-feldspar is formed by substitution of K^+ by Na^+. However, the boundary between altered and unaltered is too narrow, say only 40 nm for plagioclase, which arouses skepticism.

Another hypothesis is dissolution–precipitation (or interface-coupled dissolution–precipitation nominated by Putnis et al. 2007) by which the co-oriented albitization of feldspar is formed when hydrothermal fluid entering. Along with dissolution of an old mineral, a new replacive mineral immediately crystallizes, and this new mineral may take the same orientation as that of the old one because replacement is proceeded within the mineral.

Fig. 2.10 BSE image. **b** Magnified frame in (**a**). Micropores in K-feldspar K accompanied with intense argillation and dense albite lamellae. Huangnitian quartz syenite

(a)   **(b)**

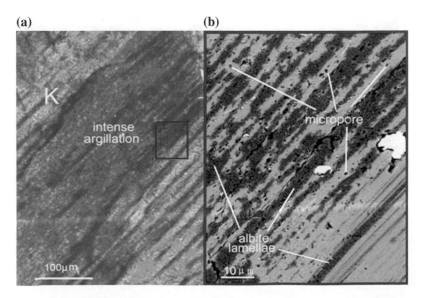

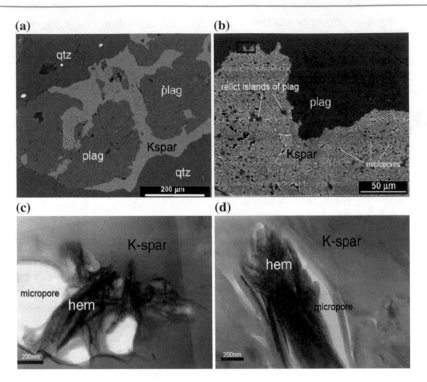

Fig. 2.11 Plagioclase is replaced by K-feldspar. **a, b**— Back-scattered SEM images; **c, d** Transmission electron microscope images. **a** A higher degree of replacement in which the potassium feldspar (*pale gray*) forms along most plagioclase grain boundaries and also transects original plagioclase single crystals, forming relict islands of crystallographically continuous plagioclase within the K-feldspar. **b** The lobate (crenulated) grain boundaries and overall texture are interpreted as a replacement of plagioclase by alkali feldspar, which contains numerous small pores (*black*), and relict islands of plagioclase (*darker gray*). **c** Micropore, containing hematite, at the interface between plagioclase and K-feldspar from the Finnish granite. **d** Hematite-filled micropore in K-feldspar from the Itapoã granite, Brasil. From Putnis et al. (2007) (With kind permission from Putnis)

For albitization of plagioclase, the more developed the micropores, the more intense is the co-oriented albitization.

As for K-feldspar, the situation is somewhat confusing. The co-oriented albitization of K-feldspar begins in the K-feldspar phase acompanied or pioneered with cloudy hydrogoethitization (Fig. 1.68) and neither along cleavage nor close to perthitic albite lamellae. The rule of developing micropores in K-feldspar is not very clear, so the understanding of the co-oriented albitization mechanism for K-feldspar is still deficient.

2.6 Relation Between the Two Patterns of Replacement

What is the relation between the two patterns of replacement?

Hetero-oriented replacement occurs at grain boundaries between two minerals, while co-oriented replacement takes place inside the replaced mineral. Obviously, it is easy to imagine that a hydrothermal gas–fluid moves first along grain boundary before entering the interior of mineral, therefore hetero-oriented replacement happens generally earlier than co-oriented replacement. Certainly, it may also infer that the two patterns of replacement might take place almost simultaneously.

The rock-forming minerals which may undergo both hetero-oriented and co-oriented replacement processes are mica and feldspar. Biotite may be either hetero- or co-orientedly replaced by muscovite. Both K-feldspar and plagioclase may be either hetero- or co-orientedly replaced by albite. Plagioclase can be hetero-orientedly replaced by K-feldspar. However, the authors doubt whether plagioclase, including albite, as well as K-feldspar may be co-orientedly replaced by K-feldspar.

2.6.1 Muscovitization of Biotite

Biotite commonly may be transformed into muscovite because of co-oriented replacement by muscovite (Figs. 1.61 and 1.62). Muscovite may also hetero-orientedly replace biotite (Figs. 1.41 and 1.42) in mild degree. They appear singly and separately. The authors have not seen where the two kinds of replacing muscovite meet together so far. The process sequence between them is not clear yet.

2.6.2 Albitization of K-Feldspar

Albitization of K-feldspar is divided into two patterns: hetero-oriented and co-oriented replacement. They may be observed in a sodium metasomatite. Then, what is the relationship between them?

As far as we know, the earliest hetero-oriented albitization of K-feldspar was formed at post-magmatic or deuteric stage and was widespread in the whole granitic body. Much later than the former, the co-oriented albitization of K-feldspar occurred due to action of intense alkali hydrothermal solutions in a local region, where at least microbroken deformation took place after solidification of multiple pluton in a batholith. However, no example of co-oriented albitization of K-feldspar in deuteric stage has been found so far. Nevertheless, the late hetero-oriented albitization may also emerge after the calcitification which postdates the co-oriented albitization of K-feldspar.

The intensity of hetero-oriented albitization might be represented by the size of newly formed albite grains, ranging from small (<0.1 mm) to large (>0.5 mm). However, no co-oriented albitization of K-feldspar has been observed even in granite with large size (up to 0.5–1 mm) of newly formed albite consisting of 15 % of the whole rock (Figs. 1.8, 2.12 and 2.13). It means that the co-oriented albitization of K-feldspar could not be initiated even if the hetero-oriented albitization of K-feldspar developed intensely.

On the other hand, in the alkali metasomatite that has been subjected to intense sodium metasomatism nearly all the primary K-feldspar crystals up to 30–45 % were transformed or co-oriented albitized to chessboard albite. Here the size of clear albite rim belonging to hetero-oriented albitization remained as the same (<0.1 mm) without broadening (Figs. 2.14 and 2.15), and the width of swapped albite rows was similar as before (Figs. 2.16 and 2.17). In other words, there is no trace of enhancing the hetero-oriented albitization after the K-feldspar was co-orientedly albitized completely to chessboard albite. So, the intense co-oriented albitization of K-feldspar also does not promote the

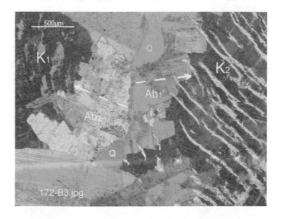

Fig. 2.12 (+)Q. No co-oriented albitization of K-feldspar was provoked by intense hetero-orientation albitization (Ab$_1$′Ab$_2$′). Naqin leucogranite

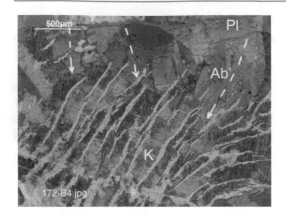

Fig. 2.13 (+)Q. No co-oriented albitization of K-feldspar was triggered by intense hetero-orientation albitization Ab'. Naqin leucogranite

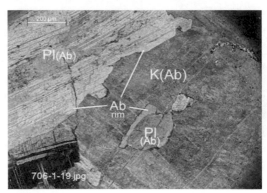

Fig. 2.15 (+). Albite clear rim keeps the same size after K-feldspar was completely co-orientedly albitized K(Ab). Sodium metasomatite in Jiling granite

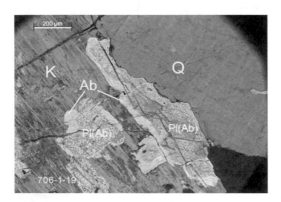

Fig. 2.14 (+). Albite clear rim in ordinary granite. Jiling granite, Gansu Province

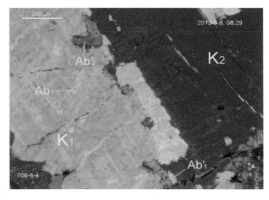

Fig. 2.16 (+)Q. Swapped albite rows between two K-feldspars in Jiling granite, Gansu Province

occurrence of the hetero-oriented albitization of K-feldspar.

These are actual facts that the authors have observed for years. We cannot see any other alternative explanation but to recognize that the processes of the two patterns of albitization took place and proceeded individually and independently, although they were all directed against K-feldspar and were created commonly by a sodic hydrothermal solution. There is no continuous or transition relation between them, although they are all generally called albitization of K-feldspar.

It may be inferred that their conditions of occurrence and the environments in which they were formed are probably quite different and they

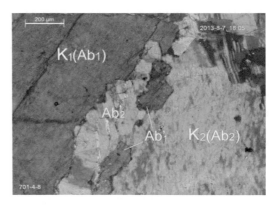

Fig. 2.17 (+)Q. Swapped albite rows keep the similar width after K-feldspar was thoroughly co-orientedly albitized K(Ab). Sodium metasomatite, Jiling granite

should reasonably be distinguished and not be confused each other.

2.6.3   Albitization of Plagioclase

In early stage during hetero-oriented albitization, K-feldspar is easily replaced while most perthitic albite will remain in situ. Only a few tiny perthitic albite accompanied with K-feldspar might be replaced simultaneously, while whole euhedral plagioclase keeps its integrity and even maintains its zonal structure. At that time, plagioclase is fresh and unsericitized (Fig. 1.3) and co-oriented albitization of plagioclase does not occur.

After later alteration due to the introduction of either hydrothermal fluid or groundwater, plagioclase is partly or wholly blurred, sericitized, and transformed into albite with decreased An value, i.e., co-orientedly albitized.

In contrast, wherever plagioclase has been subjected to intense alteration, resulting in perfect co-oriented albitization, the early-formed clear albite rim and intergranular albite keep their original widths as before. So, intense co-oriented albitization of plagioclase also does not enhance the hetero-oriented albitization.

It indicates that the hetero-oriented albitization and co-oriented albitization of plagioclase took place separately and respectively.

Therefore, the formation conditions and environments of the two replacement patterns are different and not associated with each other.

The authors emphasize that it is reasonable and necessary to divide metasomatic phenomena into the above two patterns and to describe and treat them separately, rather than to confuse them, although their individual formation mechanisms and conditions are poorly understood so far.

References

Alexander GB, Heston WM, Iler HK (1954) The solubility of amorphoussilicain water. J Phys Chem 58:453–455

Banes JO, Amourc M (1984) Biotite chloritization by interlayer brucitization as seen by HRTEM. Am Mineral 69:869–871

Behrmann JH (1985) Crystal plasticity and superplasticity in quartzite; a natural example. Tectonophysics 115:101–129

Brown WL, Parsons I (1993) Storage and release of elastic strain energy: the driving force for low temperature reactivity and alteration in alkali feldspars. In: Boland JN, FitzGerald J (eds) Defects and processes in the solid state: geoscience applications. Elsevier, pp 267–290

Carmichael DM (1986) Induced stress and secondary mass transfer: thermodynamic basis for the tendency toward constant-volume constraint in diffusion metasomatism. In: Halverson HC (ed) Chemical transport in metasomatic processes, vol 218. Nat ASI Series C, pp 237–264

Collins LG (1998) Primary microcline and myrmekite formed during progressive metamorphism and K-metasomatism of the Popple Hill gneiss, Grenville Lowlands, northwest New York, USA, Myrmekite. http://www.csun.edu/ ~ vcgeo005/Nr28Popple.pdf. ISSN 1526-5757

Collins LG, Collins BJ (2002) K-metasomatism and the origin of Ba- and inclusion-zoned orthoclase megacrysts in the Papoose Flat pluton, Inyo Mountains, California, USA, Myrmekite. http://www.csun.edu/ ~ vcgeo005/Nr44Papoose.pdf. ISSN 1526-5757

Collins LG (2003) Transition from magmatic to K-metasomatic processes in granodiorites and Pyramid Peak granite, Fallen Leaf Lake 15-Minute Quadrangle, California, Myrmekite. http://www.csun.edu/ ~ vcgeo005/Nr48Fallen.pdf. ISSN 1526-5757

Correns CG (1950) Zur Geochemie der Diagenese. I. Das Verhalten von CaCO3 und SiO2. Geochim Cosmochim Acta 1:49–54

Deer WA, Howie RA, Zussman J (1963). Rock-forming minerals. 3.4. Longmans, London

Dengler L (1976) Microcracks in crystalline rocks. In: Wenk HR (ed) Electron microscopy in mineralogy. Springer, pp 550–556

Engvik AK, Putnis A, Gerald JDF (2008) Albitization of granitic rocks: The mechanism of replacement of oligoclase by albite. Can Mineral 46:1401–1415

Farver JR, Yund RA (1995) Grain boundary diffusion of oxygen, potassium and calcium in natural and hot-pressed feldspar aggregates. Contrib Mineral Petrol 118:340–355

Heydemann A (1966) Uber die Chemiche Verwitterung von Tonmineralon (Experimentelle Untersuchungen). Geochim Cosmochim Acta 30:995–1035

Hiraga T, Nagase T, Akizuki M (1999) The structure of grain boundaries in granite-origin ultramylonite studied by high-resolution electron microscopy. Phys Chem Miner 26:617–623

Hövelmann P et al (2010) The replacement of plagioclase feldspars by albite: Observations from hydrothermal experiments. Contrib Mineral Petrol 159:43–59

Kogure T, Banfield JF (2000) New insights into the mechanism for chloritization of biotite using polytype analysis. Am Mineral 85:1202–1208

Krauskopf KB (1956) Dissolution and precipitation of silica at low temperatures. Geochim Cosmochim Acta 10:1–26

Labotka TC, Cole DV, Fayek M et al (2004) Coupled cation and oxygen-isotope exchange between alkali feldspar and aqueous chloride Solution. Am Mineral 89:1822–1825

Lee MR, Waldron KA, Parsons I (1995) Exsolution and alteration microtextures feldspar phenocrysts from the Shap granite. Mineral Mag 59:63–78

Maliva RG, Siever R (1988) Diagenetic replacement controlled by force of crystallization. Geology 16:688–691

Merino E, Dewers T (1998) Implications of replacement for reaction-transport modeling. J Hydrol 209:137–146

Merino E, Nahon D, Wang Y (1993) Kinetics and mass transfer of pseudomorphic replacement: Application to replacement of parent minerals and kaolinite by Al, Fe, and Mn oxides during weathering. Am J Sci 293:135–155

Montgomery CW, Brace WF (1975) Micropores in plagioclase. Contrib Mineral Petrol 52:17–28

Okamoto GO, Takeshi O, Katsumi G (1957) Properties of silica in water. Geochim Cosmochim Acta 12:123–132

Ostapenko GT (1976) Excess pressure on the solid phases generated by hydration (according to experimental data on hydration of periclase). Geochem Int 13(3): 120–138

Parsons I (1978) Feldspars and fluids in cooling protons. Mineral Mag 42:1–17

Putnis A (2009) Mineral replacement reactions. Rev Mineral Geochem 70:87–124

Putnis A, Hinrichs R, Putnis CV et al (2007) Hematite in porous red-clouded feldspars: evidence of large-scale crustal fluid–rock interaction. Lithos 95:10–18

Que M, Allen AR (1996) Sericitization of plagioclase in the Rosses granite complex, Co. Donegal, Ireland. Mineral Mag 60(6):927–936

Roedder E, Coombs DS (1967) Immiscibility in granitic melts, indicated by fluids inclusions in ejected granitic blocks from Ascension Island. J Petrol 8:15–17

Rong JS (1982) Microscopic study on the phenomenon of granite mineral replacement. In: Petrological Study(1). Geological Publishing House, Beijing, pp 96–109 (in Chinese)

Rong JS (2009) Two patterns of monomineral replacement in granites. Myrmekite. http://www.csun.edu/~vcgeo005/Nr55Rong3.pdf. ISSN 1526-5757

Veblen DR, Ferry JM (1983) A TEM study of the biotite-chlorite reaction and comparison with petrological observations. Am Mineral 68:1160–1168

Wang ZH, Liu R (2009) Characteristic research of micropores and microcracks in alkali feldspar. Bull Mineral Petrol Geochem 28(2):143–146 (in Chinese)

Worden RH, Walker FDL, Parsons I et al (1990) Development of microporosity, diffusion channels and deuteric coarsening in perthitic alkali feldspars. Contrib Mineral Petrol 104:507–515

Discussion About the Origin of Mineral Textures in Granite

3

3.1 Origin of Cleavelandite (Small Platy Albite)

Formed as late and small intrusion in multistage complex batholith, Li-F granite is ultra-acidic, alkali, peraluminous, generally rich in rare elements, such as Li, F, Rb, Cs, Nb, Ta, W, Sn, Be, etc. As fine valuable rare minerals, such as pyrochlore, microlite, niobite, tantalite et al., are scattered in Li-F granite, which is nominated as rare metallic mineralized granite. The typical Li-F granite is characterized by vertical zonation and divided into, from the bottom upwards, two-mica granite → muscovite granite or leucogranite → topaz-lithium (or zinnwaldite) bearing granite → albite granite → greizen → K-feldspar pegmatic and quartz shell (Zhu et al. 2002; Jun et al. 2008). From bottom to top, the composition of plagioclase is gradually changed from sodic oligoclase to albite; with albite content increased toward the top faces, the content of K-feldspar and quartz decreased; mica minerals change from lithium-biotite → protolithionite or two-micas → muscovite or lepidolite; grain size ranges from medium → medium-fine → fine.

A lot of small platy albite (cleavelandite) crystals are distributed at random (Figs. 3.1, 3.2, 3.5 and 3.6) and enclosed in bigger quartz and K-feldspar, even in topaz (Fig. 3.4), sometimes showing interzonal arrangements in large quartz crystals (called "snow ball") (Fig. 3.3). The texture of Li-F granite is quite distinctive compared with ordinary granite.

3.1.1 Metasomatic Hypothesis

These small platy albite crystals were mainly thought to be metasomatic in origin in 1960–70.

Masgutov (1960) regarded these small platy albite crystals as one of three kinds of albitization classification that he proposed; i.e., the chaotic albite replacement kind. (The other two kinds of albitization are the perthitization of K-feldspar and the deanorthitization of plagioclase).

On the basis of the hydrothermal metasomatic origin of the small platy albite, many researchers consider that Li-F granite is rich in cleavelandite and thus named it as apogranite (апограниты) is formed by postmagmatic hydrothermal alteration from normal granite (Beus et al. 1962). Beus showed a series of microphotographs of enclosures of platy albites in quartz grains, ranging from random in small quartz grains to slightly circularly enclosures in quartz of medium size, and finally to evidently concentric inclusions in quartz of large size. He believed that either these small albite crystals or the large quartz (and K-feldspar) including the albite crystals are of metasomatic origin.

Metasomatists consider the small platy albite crystals as metasomatic products perhaps due to the fine size and their special chaotic distribution pattern which are not observed in ordinary granite. They emphasize that metasomatism and alteration may expel rare metals upwards from preexisting mica and feldspar resulting in the formation of ore deposits in the upper part of a

J. Rong and F. Wang, *Metasomatic Textures in Granites*, Springer Mineralogy, DOI 10.1007/978-981-10-0666-1_3

Fig. 3.1 (+)Q. Random distribution of fine platy albite crystals in xenomorphic quartz. Yashan Li-F granite, Jiangxi Province

Fig. 3.2 (+). Lepidolite albitite is formed by aggregate of platy albite Ab with lepidolite Ms. Yashan Li-F granite

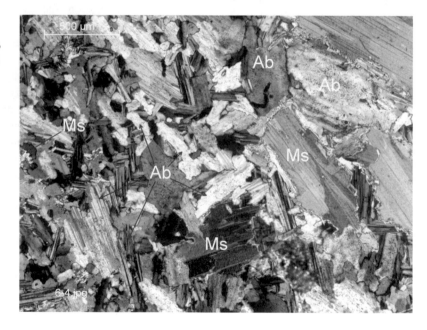

rock body (Masgutov 1960; Beus et al. 1962; Aubert 1964; Burnol 1974; Hu 1975; Wang 1975; Hong 1975; Stemprok 1979; Imeopkaria 1980; Yuan et al. 1987; Wang et al. 1989; Xia and Liang 1991).

3.1.2 Magmatic Hypothesis

Other researchers, however, insist on a magmatic origin of the Li-F granites, including the cleavelandite. Therefore, the rare metal mineralization

Fig. 3.3 (+)Q. Concentric ring of small albite crystals in a big quartz crystal (snowball texture). Yashan Li-F granite

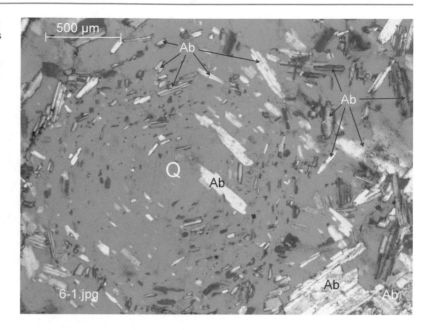

Fig. 3.4 (+)Q. Small albite crystals are enclosed in a big anhedral topaz. Yashan Li-F granite

is of primary genesis (Wang et al. 1970; Kovalenko 1971; Liu 1975; Eadington 1978; Rong 2009; Du 1984; Raimbault 1984; Zhang 1985; Cuney 1985; Xia 1989; Zhu 1992; Taylor 1992).

The authors doubt about the metasomatic origin of these platy albite. Even if they had been formed by replacement, the replacement should have a hetero-oriented pattern.

If the hetero-oriented replacement pattern proposed by the authors is accepted as a believable and ubiquitous metasomatic rule as illustrated in Figs. 1.2, 1.3, 1.4 and 1.8, quartz should be stable and cannot be replaced by albitization, and the replacive albite should occur regularly at the grain boundaries either of plagioclase with K-feldspar or between two K-feldspars rather

Fig. 3.5 (+)Q. The small albite crystals in quartz are of primary, since quartz cannot be replaced by albite. So are the albite crystals in K-feldspar, as the size of platy albite crystals in both quartz and K-feldspar is in identical scale

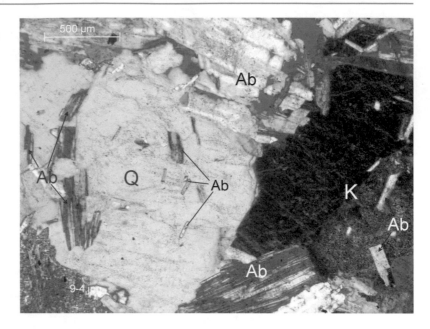

Fig. 3.6 (+)Q. No swapped albite rows are seen at the grain boundaries among various K-feldspars (K1, K2, K3, K4). Yashan Li-F granite

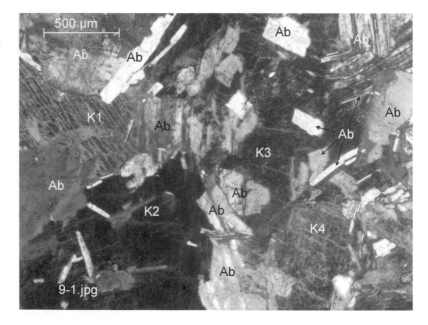

than randomly or zonally in quartz or K-feldspar. Therefore, the small platy albite grains distributed in quartz must be primary rather than metasomatic.

The following important phenomena under the microscope may be useful for the determination of their genesis.

① The small platy albite crystals are euhedral or subhedral. They are distributed chaotically and/or may be zonally arranged in quartz and K-feldspar.

② As quartz can not be replaced by albite, the small platy albite crystals in quartz must be just inclusions. K-feldspar may be replaced by albite, but now the size, form, and distribution of the small platy albite in K-feldspar are the same as those in quartz (Figs. 3.5 and 3.6), demonstrating that no notable K-feldspar is replaced by albite, i.e., no evident metasomatic growth of albite in K-feldspar. So the platy albite crystals in K-feldspar are inclusions too.

③ The fact that no distinct swapped rims of albite occur along the grain boundaries between two K-feldspars with different orientations indicate that at least the hetero-oriented albitization in the rock is very weak (Fig. 3.6).

④ Concentric enclosures of small platy albite crystals in quartz manifest that they are primary rather than metasomatic, because the metasomatic relicts can only be preserved in situ. The relicts are impossibly to be pushed aside, rearranged zonally, and even changed in their orientations. Zonal inclusions of fine guest crystals in host minerals are not an uncommon texture in igneous rocks. Examples are the concentric enclosures of accessory minerals, such as apatite, zircon, etc., in biotite (or in hornblende).

From the above mentioned, we deduce that the small platy albite grains are magmatic rather than metasomatic and the hetero-oriented albitization developed poorly at least.

The discovery of ongonite (subvolcanic phase of Li-F granite with rare earth elements, such as granite porphyry bearing topaz and quartz keratophyre), the study of fusion inclusions and experimental petrological studies have also shown that Li-F granite can directly be formed from a granitic melt rich in Li-F at the top of an intrusion rich in alkali and volatile elements during decreasing temperature. The vertical zonation of Li-F granite was even partly originated from liquid segregationmechanism as differentiation function (proposed by Wang et al. 1998; Wang and Huang 2000).

3.2   Origin of Myrmekite

Myrmekite was first described by Michel-Lévy (1874) and named by Sederholm (1897) as an intergrowth of plagioclase and quartz vermicules. Myrmekite is commonly observed in felsic-to-intermediate calc-alkalic plutonic rocks (but not in alkalic-tending rocks) and in granitic gneisses

Fig. 3.7 (+). Myrmekite (An_{14}) at contact of plagioclases (An_{28}) with K-feldspar. No myrmekite among plagioclases. Lantian granite, Shaanxi Province

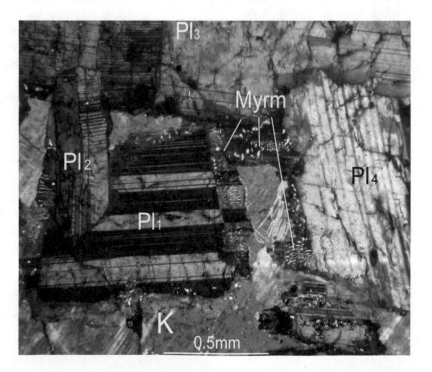

Fig. 3.8 (+). Myrmekite
appears at boundary of
plagioclase (Pl$_1$) with
K-feldspar (K$_1$, K$_2$, K$_3$).
No myrmekite at contact
with quartz. Lantian
granite, Shaanxi Province

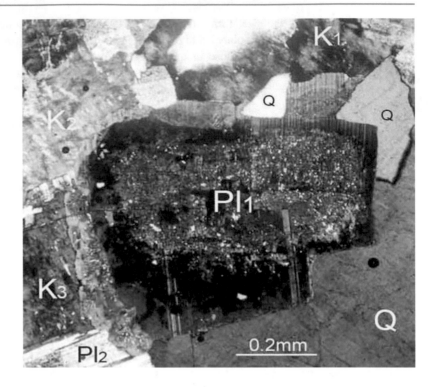

of similar composition. Myrmekite is found
along the grain boundary between plagioclase
and K-feldspar (Fig. 3.7) and is absent at the
contact between various plagioclases and plagioclase with quartz (Fig. 3.8).

On the basis of localization and occurrence,
Phillips (1974) classified myrmekite in the following ways:

① Rim myrmekite, occurring at the contact of
plagioclase with another differently oriented
K-feldspar (Figs. 3.7, 3.8);

② Intergranular myrmekite, situated at the
boundary between two differently oriented
K-feldspars (Fig. 3.17);

③ Wartlike myrmekite, located on the border
of a K-feldspar megacryst, with the curve
convex toward the K-feldspar (Fig. 3.9);

④ Enclosed   myrmekite,   in   K-feldspar
(Fig. 3.15).

Myrmekitic plagioclase has the same crystallographic orientation as that of the primary plagioclase nearby on one side, but undoubtedly
different from that of the K-feldspar nearby on
the other side.

The quartz in myrmekite is vermicular
(tapered, curved, and/or sinuous) and rod-like
with round to oval sections and elongated toward
the myrmekite border. The thickness of vermicular quartz may be homogeneous in one rock
body or vary from fine (<0.005 mm) to coarse
(>0.015 mm); thus, myrmekite may be classified
as fine myrmekite, meso-myrmekite, and coarse
myrmekite (Fig. 3.9).

Myrmekite containing coarse quartz vermicules in some places is bordered by meso-to-
fine quartz vermicules. The coarser vermicules
are commonly close to the plagioclase upon
which the myrmekite is connected, while the
myrmekite containing finer quartz vermicules is
toward the K-feldspar. It suggests that the coarser
quartz vermicules are formed earlier than the finer
quartz vermicules. Correspondingly, the plagioclase   of   myrmekite   varies   from   more
calcic to sodic. However, they may exist separately (Fig. 3.9). The volume or size of the
vermicular quartz is directly proportional to the
An content of the plagioclase in myrmekite
(Fig. 3.10).

Fig. 3.9 (+)Q. Wartlike myrmekite at the rim of K-feldspar. Coarse myrmekite (myrm 1) and fine myrmekite (myrm 2) occur separately. Liudaogou granite, Fengning County, Hebei Province

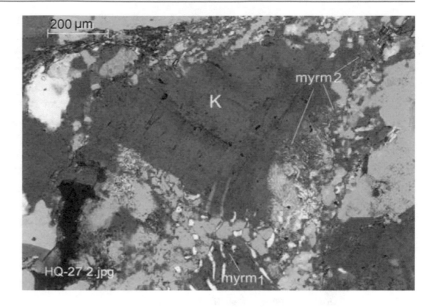

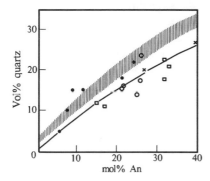

Fig. 3.10 Comparison between quartz volume % and An mol% of the host plagioclase with the theoretical relation for the strict Schwantke exsolution and Becke replacement models (*continuous line*). The hatched band shows the general observations of Becke (1908). The data points are: *dot*, Barker (1970); cross, Phillips and Ransom (1968); *circles* and *squares*, Ashworth (1972). After Smith (1974)

Generally speaking, the finer the vermicular quartz and the more the strips, the less the quartz content and the lower the An value of myrmekitic plagioclase. On the contrary, the coarser the quartz and the less the strips, the more the quartz content and the higher the An value of myrmekitic plagioclase. The An value myrmekitc plagioclase is commonly lower than that of the primary plagioclase. The highest An value of myrmekitic plagioclase may range up to that of the primary plagioclase.

3.2.1 Hypotheses for Origin of Myrmekite

There are at least five major hypotheses for the origin of myrmekite.

3.2.1.1 Replacement of Preexisting Plagioclase by K-Feldspar

Drescher-Kaden (1948) considered that myrmekite is formed from replacement of plagioclase by a fluid bearing K and Si. The crystallization of SiO_2 from fluid (bearing SiO_2) penetrating in plagioclase forms myrmekitic quartz. Myrmekite is obviously older than K-feldspar. If replacement proceeds thoroughly, the plagioclase can be replaced by K-feldspar. Based on this theory, granitizers hold that K-feldspar is produced by K-feldspathization of plagioclase.

3.2.1.2 Replacement of Preexisting K-Feldspar by Sodic Plagioclase

Becke (1908) held that myrmekite is formed by partial replacement of preexisting K-feldspar by sodic plagioclase. After the replacement of K-feldspar by sodic plagioclase, the surplus SiO_2 was retained and precipitated as vermicular quartz in the newly formed acidic plagioclase.

Four molecules of quartz are produced when each molecule of anorthite is introduced.

$$2KAlSi_3O_8 + 2Na^+ \rightarrow 2NaAlSi_3O_8 + 2K^+$$

$$2KAlSi_3O_8 + Ca^{2+} \rightarrow CaAl_2Si_2O_8 + 4SiO_2 + 2K^+$$

According to "Rock-Forming Minerals" (Deer et al. 1963), the SiO_2 content of anorthite is about 44–46 %, while that of K-feldspar and albite are 64–67 %, and 65–68 %, respectively. The SiO_2 content of anorthite is really less than that of K-feldspar and albite.

3.2.1.3 Exsolution or Unmixing of K-Feldspar

Schwantke (1909) suggested that the K-feldspar crystallized from magma is formed by solid solution, which consists of major K-feldspar $KAlSi_3O_8$, minor albite $NaAlSi_3O_8$, and a few anorthite $CaAl_2Si_2O_8$. He postulated that both $NaAlSi_3O_8$ and so-called Schwantke molecule [$CaAl_2Si_6O_{16}$ or $Ca(AlSi_3O_8)_2$] are supposedly contained in a former high-temperature K-feldspar (Schwantke 1909; Spencer 1945). With decrease in temperature the exsolution of $NaAlSi_3O_8$ and $CaAl_2Si_6O_{16}$ occurs. The unmixing of Schwantke molecule results in the formation of myrmekite as each Schwantke molecule may exsolve $CaAl_2Si_2O_8$ and $4SiO_2$. The exsolved $CaAl_2Si_2O_8$ combined with the exsolved $NaAlSi_3O_8$ forms plagioclase and $4SiO_2$ and remains as vermicular quartz, thus forming myrmekite. The inner diffusion drives the exsolved material into the plagioclase nearby.

3.2.1.4 Recrystallization of Plagioclase

Collins (1988) considered that after cataclastic deformation and prior to introduction of K-bearing fluids, plagioclase was altered by hydrothermal solutions. In local places in the altered plagioclase lattice, the loss of Ca and Al and the retention of Na are accompanied by an increase in excess silica, resulting in the formation of quartz vermicules in myrmekite when these places recrystallized. The An content of plagioclase of myrmekite is about half the An content of the primary plagioclase.

3.2.1.5 Complex Hypothesis

Complex hypothesis were promoted by Ashworth (1972) and Phillips (1974). They considered that myrmekite may be formed by either exsolution or replacement. Exsolution origin is appropriate for the myrmekite in undeformed high level intrusions, such as rim myrmekite, grain boundary myrmekite, and general enclosed myrmekite. However, the exsolution origin does not fit for some myrmekite in deformed metamorphic rocks. The wartlike myrmekite (especially bigger myrmekite appeared in smaller K-feldspar) should be explained by replacement origin.

3.2.2 Discussion of the Origin of Myrmekite

Myrmekite is characterized by stable correlation of the volume percent of vermicular quartz with the An content of myrmekitic plagioclase (Fig. 3.10). The correlation is obviously not accidental, although precise measurements of the volume percent of vermicular quartz are difficult. Most hypotheses are consistent with this correlation.

Drescher-Kaden (1948) concluded that some vermicular quartz in K-feldspar may be relicts of replaced myrmekite, occurring as "ghost myrmekite" (Figs. 3.11 and 3.12). This relationship might serve as tenable evidence indicating that K-feldspar replaces the plagioclase of former myrmekite, and therefore, the whole K-feldspar must be metasomatic in origin. This phenomenon, however, according to the authors' point of view, can be explained by superimposition of myrmekite by the reverse K-metasomatism, i.e., K-feldspathization, which took place after the formation of myrmekite had ceased. The newly formed K-feldspar of hetero-orientation replacement pattern grew on the back preexisting K-feldspar, taking its crystallographic orientation and replacing the myrmekite in front. In this way, the myrmekite was locally destroyed by the second reverse K-feldspathization. The quartz vermicules of myrmekite remained as relicts because they were barely replaced by K-feldspathization.

Fig. 3.11 (+). Ghost myrmekite is formed by superimposition of myrmekite by the follow-up (sequential) K-feldspar replacement

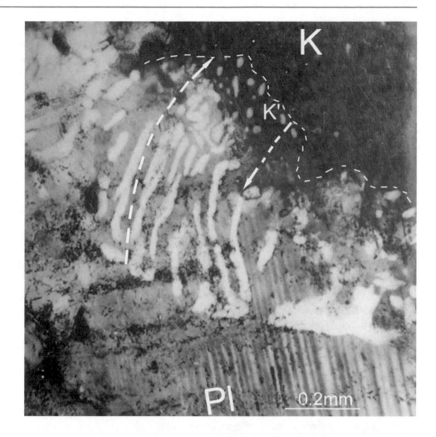

Fig. 3.12 (+)Q. Ghost myrmekite. The myrmekitic plagioclase is thoroughly replaced by the replacive K-feldspar K′ with residual myrmekitic quartz

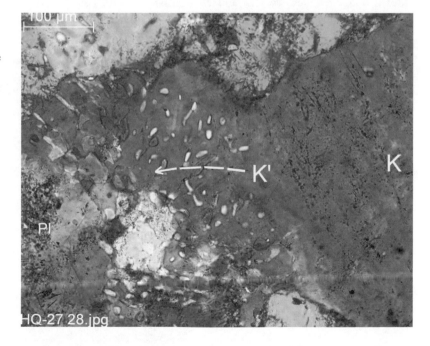

The hypothesis of "recrystallization of altered plagioclase" does not seem to apply to granitic rocks containing myrmekite that the authors have studied for the following reasons:

(1) The shape of myrmekite does not look like primary plagioclase (Figs. 3.8 and 3.9), but only seems to be similar in some places.

(2) The granitic rocks containing myrmekite are not necessarily affected by deformation, alteration, and recrystallization.

(3) Myrmekite is not formed during saussuritization and sericitization.

(4) Myrmekite may occur regularly in swapped grains at the contacts between two K-feldspar crystals.

(5) The An value of myrmekitic plagioclase may not necessarily be equal to the half An value of primary plagioclase, but is generally lower than that of primary plagioclase.

Exsolution from solid solution of alkali feldspar is an attractive hypothesis because no penetration of fluid material is needed from outside. That model is consistent with the idea of those researchers who do not believe or consider that such compact granitic rocks might have been entered and circulated by Ca- and Na-bearing fluid from outside.

The origin of perthitic albite in K-feldspar is also commonly explained by exsolution. The crystallographic orientation of both guest perthitic albite and host K-feldspar coincides with each other consistently and unexceptionally. The orientation of myrmekitic plagioclase is certainly different from that of the host K-feldspar, but identical with that of the plagioclase against which the myrmekite abuts. Moreover, no myrmekite occurs along the border of the plagioclase (Pl_2) with the co-oriented K-feldspar (Figs. 3.13 and 3.14), when myrmekite evidently develops at contact with differently oriented plagioclase.

It is also hard to explain why perthitic albite possesses the same orientation as that of the K-feldspar, while myrmekitic plagioclase has a different orientation from the K-feldspar, if both of them were exsolved from the same host K-feldspar.

As for the presence of the Schwantke molecule, it is just a hypothesis which has not been proved yet. In addition, in some places a large size of myrmekite that occurs in a small K-feldspar crystal (Fig. 3.15) makes the exsolution origin impossible.

The fact that each row of the intergranular myrmekite has the same orientation as that of the

Fig. 3.13 (+)Q. No myrmekite appears at the border of a co-orentedly enclosed plagioclase Pl_1 in host K-feldspar, while myrmekite surely presents at the contact of differently oriented plagioclase Pl_2. Lantian granite, Shaanxi Province

Fig. 3.14 (+)Q. No myrmekite is present at the epitaxial border of plagioclase Pl$_1$ with K, when myrmekite clearly develops at contact with differently oriented plagioclase Pl$_2$. Lantian granite, Shaanxi Province

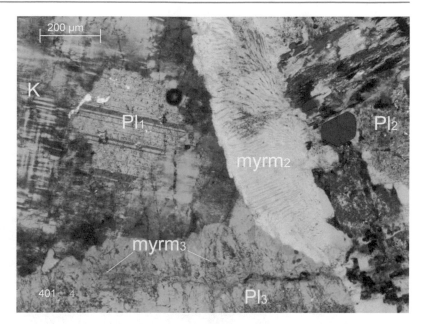

Fig. 3.15 (+)Q. A big myrmekite occurs in a smaller K-feldspar. Liudaogou granite, Fengning County, Hebei Province

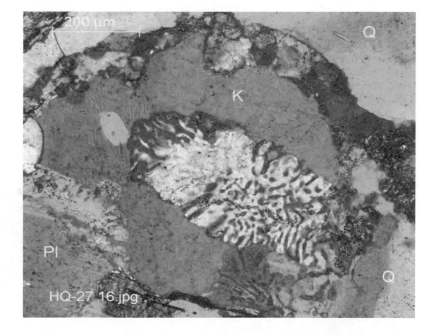

opposite K-feldspar strongly indicates that myrmekite is formed by hetero-oriented or nibble replacement. Both adjoining K-feldspars herein served as either the substrate mineral for myrmekite growth or the replaced mineral partly dissolved by Na-bearing gas–liquid involving calcium.

The authors have noticed the existence of swapped rows of myrmekite (Figs. 3.16, 3.17, 3.18, 3.19 and 3.20) on the boundary between two K-feldspar crystals in undeformed granites Rong (1982, 2002).

Some petrologists do not believe that myrmekite is formed by metasomatism, because

Fig. 3.16 Swapped myrmekitic grains occur at contact between two K-feldspars. Each row of myrmekitic plagioclase has the same orientation as that of the opposite K-feldspar, which is clearly seen when quartz plate is inserted (in *upper right* figure B)

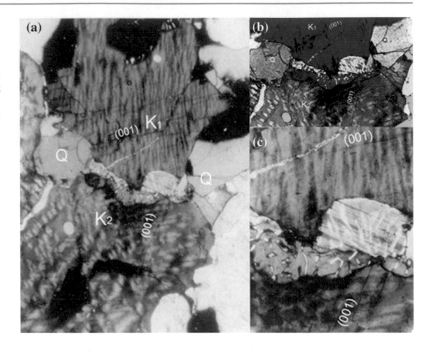

Fig. 3.17 Q. Swapped myrmkites occur at the boundary of two differently oriented K-feldspars. The replacive albite Ab$_2'$ further grew upon Myrm2' toward K$_1$. Lantian granite, Shaanxi Province

metasomatic residues of replaced perthitic albite are generally absent in myrmekite. The authors consider, at first, myrmekite has more capability of replacing the perthitic albite, as compared with replacive albite. The greater the An value of myrmekitic plagioclase, the more the capability of replacing the perthitic albite. In addition, myrmekite commonly occurs in intermediate granitic rocks, such as monzogranites, granodiorites, where the perthitic albite in K-feldspar is fine and poorly developed. Therefore, the chance for contact of myrmekite with perthitic albite is much less than that of replacive albite with perthitic albite in normal granites and alkali leucogranites. However, the relicts of perthitic albite are still seldom noticed in myrmekite (Fig. 3.19).

Fig. 3.18 (+)Q. Swapped myrmekites occur at contact between diferently oriented K-feldspars. No myrmekite appears at the border of Pl$_2$ with K$_2$, since they have the same orientation. Lantian granite, Shaanxi Province

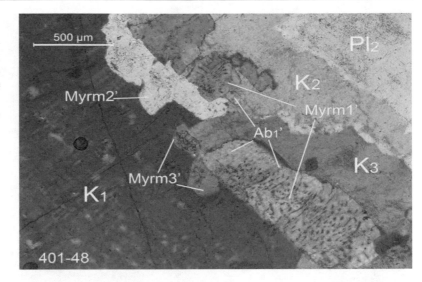

Fig. 3.19 (+)Q. Since the metasomatic ability of myrmekitization is stronger than that of albitization, relics of perthitic albite are generally hardly preserved in myrmekite. However, the relics are still found in myrmekite, when the myrmekitic plagioclase has lower An value

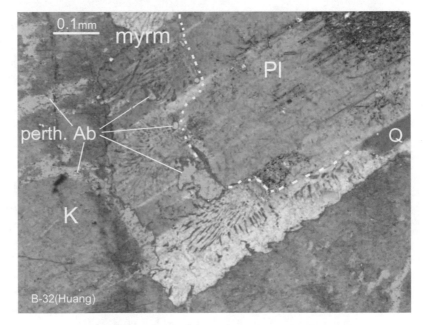

Figure 3.20 shows that two myrmekites, Myrm1′ and Myrm2′, are situated on the contact between two differently oriented K-feldspsars K$_1$ and K$_2$. The Myrm1′ and Myrm2′ are swapped myrmekites too. Moreover, there is a metasomatic relic of K$_1$ enclosed in Myrm2′. The relic maintains its original orientation of K$_1$. It precisely proves that myrmekite is reliably formed by metasomatic process.

The right coarse myrmekite Myrm2′ (D in Fig. 3.20) is somewhat special and may be subdivided into three parts on the basis of the sizes of the quartz vermicules: upper coarse myrmekite, lower fine myrmekite, and lower right quartz-tree albite. They have subtle differences in interference color, indicating the differences in composition and showing a complex sequence in which the different kinds of myrmekite were

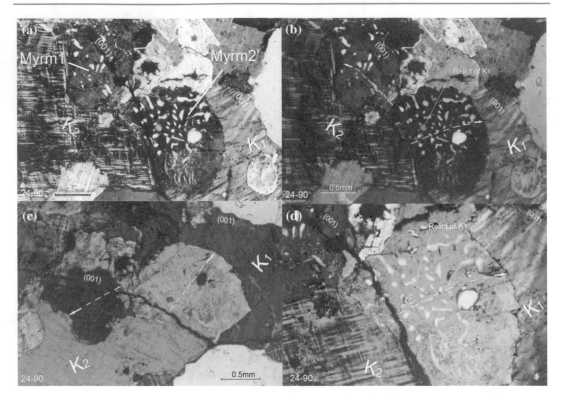

Fig. 3.20 A (+); B, C, D (+)Q. Coarse swapped myrmekite grains occur at the contact between two differently oriented K-feldspars K_1K_2. The (001) of myrmekitic plagioclase Myrm1' is parallel to that of K_1, and Myrm2' has close orientation to that of perthitic albite of K_2. A metasomatic relict of K_1 is noticeably enclosed in Myrm2'. Langshan granite, Neimeng Autonomous Region

formed, i.e., at first, upper coarse myrmekite, then, lower fine myrmekite, and finally, lower right albite. It reflects that calcium content in the sodium-bearing fluids varied from more to less and finally to zero.

The plagioclase or K-feldspar against which a myrmekite grain abuts may not be necessarily cut by a thin section, when the size of the myrmekite is large enough (Figs. 3.15 and 3.21), as if myrmekite may independently occur at the side of quartz or any other mineral.

In an extreme case, when the growth of myrmekite is vigorous in tonalite or plagiogranite with minor amount of K-feldspar, all the K-feldspar crystals may be thoroughly eliminated or replaced by myrmekite, as if myrmekite can develop in a rock which was originally free of K-feldspar.

Many small-sized myrmekite grains are commonly formed around the border of a K-feldspar (Fig. 3.9) in granite gneiss that has undergone dynamic metamorphism. However, there is no obvious plagioclase against which the myrmekite abuts.

An outstanding example of a myrmekite corona occurs as a rim on a residual rounded K-feldspar porphyroclast (Fig. 3.22) in ultramylonite, Cima di Vila, Eastern Alps, Italy (Cesare et al. 2002).

The fact that the myrmekite corona has not been affected by mylonitization shows that myrmekite is formed after mylonitization. Apparently, as mylonitization proceeded, around the K-feldspar porphyroclast, there must have been cataclastic debris of various minerals including feldspars, which obviously served as nuclei for the growth of myrmekite. Since the matrix of mylonite was completely not appropriate for the growth of myrmekite, so myrmekite

Fig. 3.21 (+)Q. Myrmekite with thickness more than 0.8 mm. The plagioclase on which the myrmekite abuts is not cut by the thin section

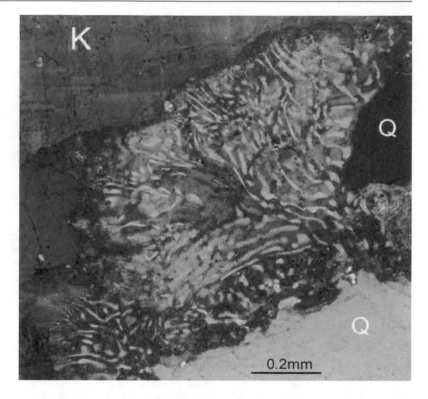

Fig. 3.22 (+). Residual rounded K-feldspar porphyroclast in ultramylonite, Cima di Vila, Eastern Alps, Italy. K-feldspar porphyroclast is rimmed by a continuous, about 1 mm thick corona of myrmekite (Myr). From Cesare et al. 2002. (With kind permission from Cesare)

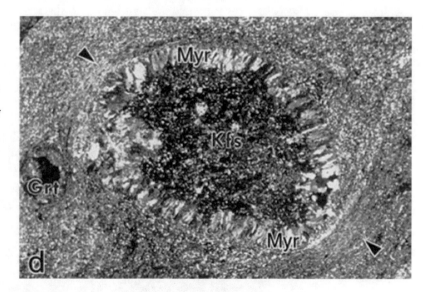

grew only around the border of the residual K-feldspar inwardly. Therefore, myrmekite is not formed by the replacement of plagioclase by K-feldspar, but on the contrary, by the hetero-oriented replacement of K-feldspar by plagioclase.

3.3 Origin of Perthite

K-feldspar crystals in granitoids rocks consist mainly of K-feldspar perthite which is normally accompanied by sodic plagioclase (albite) in

Fig. 3.23 (+). Part of albite lamellae in microcline perthite is zoned. Lantian granite

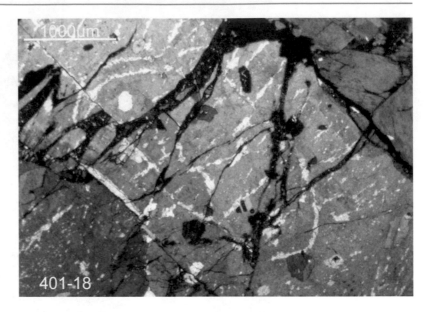

patch or lamellae form (Figs. 3.23, 3.24, 3.25, 3.26, 3.27, 3.28 and 3.29).

Perthite can be named alkali feldspar, when albite lamellae are quite abundant (Fig. 3.25). When a few blebs or patches of K-feldspar are contained in sodic plagioclase, the latter may be nominated as antiperthite (Fig. 3.42).

3.3.1   Content and Shape of Perthitic Albite

The content of albite lamellae in perthite varies with the composition of rock type. Generally the perthitic lamellae are prominent and abundant in rocks rich in Si and alkali, and poor in Ca, Fe,

Fig. 3.24 (+). K-feldspar perthite. Section intersects albite lamellae at a small angle or nearly parallel to (100)

Fig. 3.25 (+)Q. The content of perthitic albite is almost equal to or even surpasses that of K-feldspar. Langshan granite, Neimeng Autonomous Region

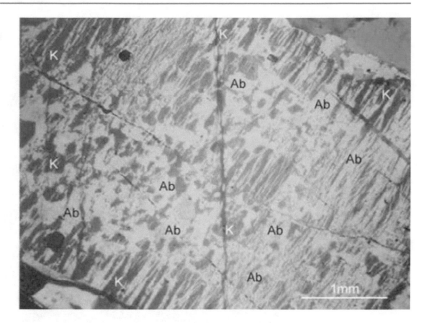

Fig. 3.26 (+). Section is ⊥axis *b*. Intersection angle of albite lamellae with (001) is about 64–73°. Albite lamellae are distributed closely parallel to the projection *N*m' of *N*m of albite lamellae or along Murchison plane ($\overline{1}502$). Naqin leucocratic granite

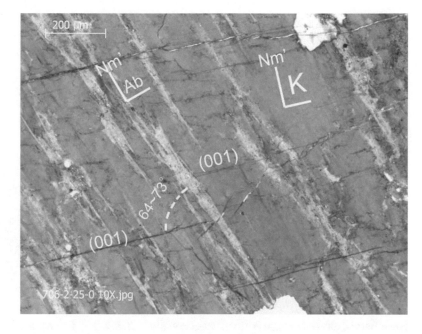

and Mg. The volumetric percentage of perthitic lamellae is generally 5–10 %, and may be as much as 30–40 %, even surpass the content of K-feldspar (Fig. 3.25). The perthitic lamellae are much less in plutonic rocks rich in Ca, Fe, and Mg and poor in Si and alkali.

K-feldspar phenocryst may contain more albite lamellae than the K-feldspar in ground mass.

Also, within a megacryst K-feldspar, the content of perthitic albite may vary gradually or abruptly from core to outer rim or vise versa.

Fig. 3.27 (+). Vein-like perthitic albite lamella are seemingly cutting across the twin composition plane (010). Naqin granite, Taishan County, Guangdong Province

Fig. 3.28 (+). Manebach twin of perthite. Perthtic albite lamellae evidently do not cut across (001). Naqin granite

Even part of albite lamellae is distributed zonally (Fig. 3.23).

The content of perthitic albite in phenocryst K-feldspar (sanidine) of a hypabyssal granite porphyry or quartz porphyry, especially effusive rhyolite, is generally much less than that in K-feldspar (orthoclase or microcline) of a plutonic granite with the same composition.

3.3.2 Form and Distribution of Perthitic Albite

Alling (1938) depicted various types of perthite as stringlets, strings, rods, beads, films, patch, flame, vein, braid, etc.

Perthitic albite lamellae are generally distributed parallel to axis *b* and close to (100),

Fig. 3.29 (+). Baveno twin of perthite. Perthitic albite lamellae evidently do not cut across the composition plane of Baveno twin

intersecting 64–73° with cleavage (001), i.e., along Murchison plane ($\overline{6}$01), ($\overline{7}$01), or ($\overline{15}$02), which are best seen in thin section //(010) (Fig. 3.26). The elongation of the major lamellae is nearly parallel to the projection of Nm of perthitic albite on (010).

In addition, perthitic lamellae are also oriented parallel to (110), (1$\overline{1}$0) forming a braided texture and rarely to (010) (Tschermark 1864).

K-feldspar (either orthoclase or microcline) and perthitic albite have the common cleavage (001). The albite twin of coarse perthitic albite is consistently connected with the cross-hatched twin of host microcline when a section is cut ⊥(010). So they have basically the same crystallographic orientation.

Additionally, the form and distribution of perthitic albite appear clearly and distinctly when the cross-hatched twinning of microcline disappears or its interference color becomes homogeneous by rotating the objective stage to a certain place under crossed polars. The optical orientation (Np, Nm, Ng) of perthitic albite is always approximately close to that of the host K-feldspar, no matter what form the perthitic albite is or how it develops. The intersection

angle of the extinction direction between them is always less than 25°, as "tracking extinction" as named by the senior author in "Preliminary Reaerch on Nanling Intrusions" (1959).

3.3.3 Hypotheses for Origin of Perthitic Albite

There are several hypotheses for the origin of perthitic albite.

(1) Contraction cracks and replacement;
(2) Exsolution from solid solution of K-Na feldspar;
(3) Simultaneous growth.

Some authors consider that there might be perthite of various origins.

3.3.3.1 Contraction Cracks and Replacement

Lehmann (1885) suggested that contraction cracks in K-feldspar allow the hydrothermal fluid to enter from outside, corrode, and replace the K-feldspar, resulting in the formation of vein-like perthite. So, the vein-like or branch-like perthite and flame perthite may be formed by contraction cracks and replacement origin.

(1) **Vein-like or branch-like perthitic albite**

The vein-like and branch-like perthitic albite occurs either irregularly or homogeneously in perthite.

① **Contraction crack and injection origin**

Andersen (1928) considered that the direction of maximum expansion lies in the (010) at +18–20° to the axis a, while the direction of minimum expansion is usually along axis b. Upon cooling the phenocryst K-feldspars could crack in a Murchison direction. He estimates that the total differential contraction from 1000 to 0 °C amounts to about 1 %. Taking the observed values of 1–20 veins of macro-albite per 2 mm in microcline perthite, the original cracks might have been only 20–1 μm wide per crack, respectively. Obviously, the calculated size of contraction cracks is much finer than that of perthitic albite in microcline. Therefore, Andersen following Lehmann postulated corrosion of microcline by solutions prior to deposition of albite in the thickened cracks (Smith 1974, p. 497).

Some researchers (Hu 1980) wrote that perthitic albite develops and distributes along cleavage and cracks of K-feldspar in a general terms. However, in fact, no perthitic albite develops and distributes along the cleavage and crack clearly seen in perthite.

Vein-like albite lamellae are commonly regarded as having injection and replacement origins by many researchers, particularly when the perthitic albite lamellae look as if they normally penetrate the Carlsbad twin composition plane (010) of the host K-feldspar (Fig. 3.27).

However, the authors consider that the phenomenon appears because the perthitic albite lamellae plane is normal to (010) and intersects at a small angle (only 2–6°, <10°) to (100). So the angle between the two perthitic albite lamellae of both sides of Carlsbad twin is less than 5–20°. Furthermore, the optical orientations of two perthitic albite lamellae on both sides of the twin are quite similar if the section cuts nearly ⊥axis c, though still subtly different with careful observation. In case of Manebach twin or Baveno twin, the idea that perthitic albite lamellae penetrate the compositional plane would naturally disappear (Figs. 3.28 and 3.29).

Therefore the consideration that perthitic albite is formed as a vein is unreliable and probably a wrong impression.

② **Comparison between perthitic albite and co-oriented albitization**

The phenomenon that the polysynthetic twinning of perthitic albite epitaxially coincides with the grid twinning of the host K-feldspar gives a strong impression that the perthitic albite is formed by replacement along some quasi-parallel microfissures which might have coalesced later on.

As above mentioned, with intense albitization, K-feldspar may be replaced during co-orientation albitization, resulting in the transformation of K-feldspar into pseudomorphic chessboard albite. Surely it does not mean that perthite is formed by co-orientation albitization?

The differences of distribution and morphological features between the perthitic albite lamellae and the co-orientedly albitized albite of K-feldspar may be compared as follows:

(a) The perthitic albite lamellae in K-feldspar are pervasively present in plutonic rocks and are not limited in some local region where intense alkali metasomatism occurs.

(b) Perthitic albite lamellae are mainly distributed along Murchison plane $(\bar{6}01)$, $(\bar{7}01)$, or $(\bar{1}502)$, intersecting at 64–73° with cleavage (001). The co-orientedly albitized albite demonstrates a somewhat different situation. The albite starts growing in forms of grouped dots, then slice (///(010)) and then in larger masses (bulks or lumps) (Figs. 1.68 and 1.69) and finally, whole K-feldspar is completely transformed into co-orientedly albitized albite.

When section is closely ⊥axis b (without albite twin), the elongation of perthitic lamellae is positive (close to Nm), while the slice of co-orientedly albitized albite is amorphous.

When section is nearly ⊥(010) (with clear albite twin), the elongation of perthitic lamellae is positive too (close to Ng'), while

Table 3.1 Microprobe analysis of perthitic albite

Perthitic albite	Mean	SiO$_2$	Al$_2$O$_3$	CaO	Na$_2$O	K$_2$O	Total	An
H$_4$	10	67.6	19.78	0.98	10.96	0.29	99.6	4.7
D$_2$	2	69.78	18.72	0.45	10.8	0.14	99.98	2.3
CL$_3$	3	67.58	20.59	3.15	8.32	0.15	99.79	17.4

H$_4$-Huangnitian quartz syenite, Yangjiang County, Guangdong Province; D$_2$-Da-ao granite, Yangjiang County, Guangdong Province; CL$_3$-Zhaojiayao granite, Xuanhua County, Hebei Province

the elongation of the slice of co-orientedly albitized albite is negative (close to Np′) (Figs. 1.70 and 1.71).

(c) Perthitic albite lamellae occur in dense polysynthetic albite twins, whereas co-orientedly albitized albite is characterized by chessboard twinning.

(d) The An value of perthitic albite varies from albite to sodic oligoclase (Table 3.1), whereas that of co-orientedly albitized albite is close to An = 1 (pure albite) (Table 1.6).

(e) Vein-like or branch-like perthitic lamellae may be replaced by either postmagmatic clear rim or intergranular albite (Figs. 1.6, 1.7 and 1.8), and both the two latters are formed earlier than the co-orientedly albitized albite (see Figs. 3.63 and 3.64). So the perthitic albite lamellae are formed much earlier than the co-orientedly albitized albite.

From the above-mentioned (a)–(e) paragraphs, we can see that the perthitic albite lamellae are quite different from the co-oriented albitized albite of K-feldspar in respect to the occurrence, distribution, form, twin features, formation sequence, etc. Therefore, we can not deduce that the vein-like or branch-like perthitic albite lamellae are formed by co-oriented albitization, although K-feldspar may really be co-orientedly replaced by albite

(2) **Flame perthitic albite**

This is a special kind of perthitic albite, which was occurring in K-feldspar of deformed granites in the greenschist facies. Flame perthitic albite has special shape and variable distribution (Figs. 3.30, 3.31 and 3.32), merging irregularly at the boundary or corner of microcline and extending inward, then thinning out or combining with other perthitic albite.

Fig. 3.30 (+). Flame perthitic albite in K-feldspar. Middle Tianshan gneissic granite, Xinjiang Autonomous Region

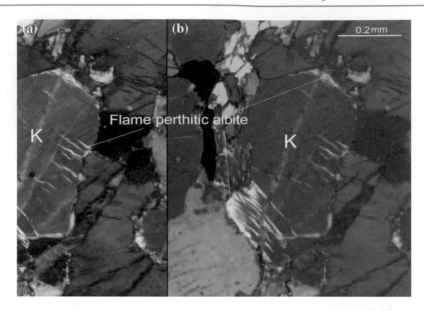

Fig. 3.31 A(+); B(+)Q. Flame perthite occurs at the corner and rim of K-feldspar. Middle Tianshan gneissic granite, Xinjiang Autonomous Region

Fig. 3.32 (+). Thin,parallel lamellae are from exsolution. Less-regular wider lamellae, abundant at the bottom and right rim are flame perthitic albite, which have the same optical orientation as albitic rims The flames are very variable in abundance, so much, so that the bottom end of the original microperthite grain is largely albite. Base of photo = 1.75 mm. From Vernon (1999), by permission

It has the same crystallographic orientation as that of the host K-feldspar and is always in a minor part. Its contact line with the host K-feldspar is clear and smooth.

Pryer and Robin (1996) suggested a replacement hypothesis for the origin of flame perthitic albite. In their model during alteration of plagioclase to albite under greenschist facies condition, a small part of the Na is liberated and migrates to the K-feldspar, then substitutes for the K, forming albite flames.

The authors have not studied flame perthite, but think that it is possibly originated by injection into cracks or by replacement. However, the forms and distributions of flame albite are obvious different from that produced by co-oriented albitization mentioned above.

3.3.3.2 Exsolution from Solid Solution of K-Na Feldspar

The solid-state exsolution theory for perthite formation began with Vogt (1905). He and subsequent workers pointed out that the homogeneously Na-bearing sanidine would gradually transform to orthoclase and microcline during slow cooling and react with hydrothermal fluid, resulting in the formation of patch perthite (at 450 °C) and straight lamellar film perthite (at ≤ 350 °C) (Parson and Lee 2009).

The fact that the perthitic lamellae are mainly distributed along $(\bar{6}01)$, $(\bar{7}01)$, or $(\bar{1}\bar{5}02)$, i.e., the Murchison plane has caused many researchers to postulate various formation mechanisms, such as contraction cracks produced by differential thermal expansion (Andersen 1928), anisotropic diffusion (Rosengvist 1950) and differences of lattice strain (Smith 1961).

Cahn (1968) proposed that the orientation of the exsolution boundary of unmixing component in the solid-state should be along the plane of the minimum value of elastic strain energy or of surface energy.

Willaime and Brown (1974, 1985) calculated the minimum value of elastic strain energy on the basis of the data for the lattice coefficient of K-feldspar and albite and concluded that the direction of the exsolution phase albite in solid-state feldspar should be basically along $(\bar{5}.\bar{7}01)$ to $(\bar{7}.\bar{9}01)$, which is very close to $(\bar{6}01)$, $(\bar{7}01)$, or $(\bar{1}\bar{5}02)$, i.e., Murchison plane, and subordinately along $(\bar{1}10)(\bar{6}31)(\bar{6}61)(\bar{6}01)$.

If the conclusion proposed by Willaime and Brown is reliable and believable, it would provide an excellent reason for accepting the solid-state exsolution theory on the formation of perthite.

The authors did microprobe scanning across two coarse (80 μm thick) perthitic albite lamellae in a K-feldspar (Fig. 3.33).

The results show that the contents of Na and K in K-feldspar are comparatively stable, and there is only a subtle decrease of Na content at the nearest outside border of the macroperthitic lamellae. If the albite lamellae are formed by unmixing, the unmixing should be produced by the whole K-feldspar rather than by the

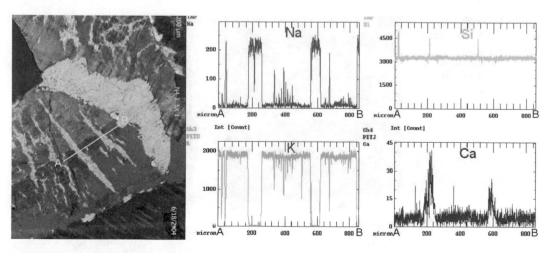

Fig. 3.33 Microprobe scanning analysis across macroperthitic albite lamellae. There is no evident decrease of Na content at the outside border of the macroperthitic lamellae in K-feldspar. The contents of Na, K, Ca are comparably stable in K-feldspar. Ca content of the macroperthitic albite is higher in the center than in the rim sides. Naqin granite

K-feldspar near the macroperthitic lamellae. However, the authors wonder whether the fact that Ca content in the center of the coarse lamellae is higher than that in the borders of the rim shows that the Ca content of the albite exsolved earlier (in the center) is higher than that exsolved later (in the rim)?

If the coarse lamellae are not generated from unmixing of solid solution, then the possibility of simultaneous growth origin has to be considered.

3.3.3.3 Simultaneous Growth

The theory of "simultaneous growth" was also proposed by Vogt (1905). He also suggested that patch perthite may be formed by simultaneous crystallization from magmatic melt. In other words, a part of albitic component is crystallized contemporaneously with K-feldspar rather than as sodic plagioclase separately.

Is it possible that perthite may be directly or contemporaneously crystallized from melt?

Lofgren (1977) successfully conducted a laboratory experiment to produce alkali feldspar, which shows that the plagioclase ($An_{25-15}Or_{5-20}$)-sanidine ($Or_{75-50}An_{2-7}$) intergrowths were directly crystallized from ternary feldspar melts. He said "The plagioclase occurs as thin, parallel lamellae of plagioclase quite regular in the (100) direction that resemble microperthite, or in patches that

resemble patch perthite. ….The compositions, the presence of zoning, and the short run times indicate that the intergrowths grew directly from the melt and did not develop by exsolution". The authors consider that the lamellae of plagioclase in the (100) direction pointed out by Lofgren may probably just be the <u>Murchison plane</u>, because the latter is close to the (100) and the experimental products are too small to be accurately measured.

3.3.4   Elongated Enclosed Crystals May Be Parallelly Distributed in Feldspar

In K-feldspar megacrysts from plutonic granitic rocks, mineral inclusions are commonly enclosed, such as plagioclase, quartz, biotite, etc. They may be arranged chaotically, if they are in isometric form (Fig. 3.34).

If the inclusions in K-feldspar are columnar, pencil like, or stick like, except disorderly arranged, they may parallelly or zonally be distributed, which is evidently seen along the section ∥(010) of K-feldspar.

Smith J V in his monograph "Feldspar minerals" Volume II (1974) (page 460) reported "Needle-shaped inclusions of apatite (luminescing yellow) were found to be common and

Fig. 3.34 (+). Random distribution of isometric crystals of plagioclase enclosed in K-feldspar Section is along (010)

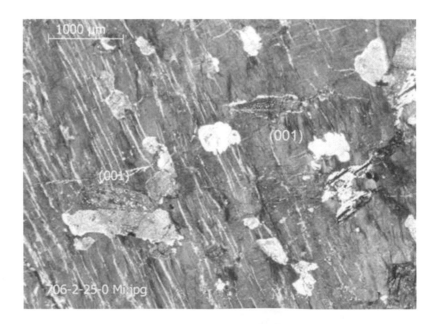

concentrated preferentially in vein albite (Smith and Stenstrom 1965). The apatite crystals tended to be needle shaped up to 1 μm across and 100 μm long. The needles tended to align in a Murchison direction".

While observing thin sections collected from granitoids from a batholith along the southwestern coastal region of Guangdong Province in 1964, the senior author was surprised to notice that acicular crystals of apatite (3 μm wide and >300 μm long) were parallelly enclosed in the primary albite (An_{4-10}) in the Shanbei leucogranite, Taishan County (Fig. 3.35) (Rong 2009).

The long acicular inclusions are arranged at an angle of about 65° to the basal cleavage (001) and approximately parallel to the projection Nm′ of Nm of albite on (010) plane. The distribution pattern of acicular apatite in albite of Manebach twin is similar to that of albite lamellae in perthite (Fig. 3.36).

This similar correlation implies that it is helpful to explain the origin of perthite.

The phenomenon that the host K-feldspar may also contain fine pencil like albite crystals (Fig. 3.37) was also observed by the senior author in 1975 when he was studying the Chengkou granite, Zhuguangshan Batholith, Guangdong Province.

Although several albite crystals are enclosed in a disordered arrangement, some are parallel to the perthitic albite.

Later, the senior author again noticed that acicular apatite crystals are also accidentally contained in K-feldspar in Rössing pegmatoid alaskitic granite, Namibia (Figs. 3.38 and 3.39). Several coarse acicular apatite crystals (5 μm wide and 400 μm long) are oriented parallel to perthitic lamellae, intersecting at 65° with (001), and also close to the projection Nm′ of perthitic lamellae, However many fine as well as some coarse apatite crystals are still enclosed in disordered arrangements (Fig. 3.39b).

The acicular apatite in feldspars and fine pencil-like albite in K-feldspars are evidently not formed by exsolution, but just are ordinary inclusions.

It is hard to know the distribution and orientation of the acicular minerals along the Murchison plane, because it is difficult to obtain a thin section along the Murchison plane. However, it is ascertained at least that their directional arrangements may occur along the intersection line of the Murchison plane with (010) plane as above mentioned.

Fig. 3.35 (+). Section is along (010). Directional inclusions of acicular apatite Ap in albite. Some short apatite inclusions are disorderly arranged. Shanbei granite, Taishan County, Guangdong Province

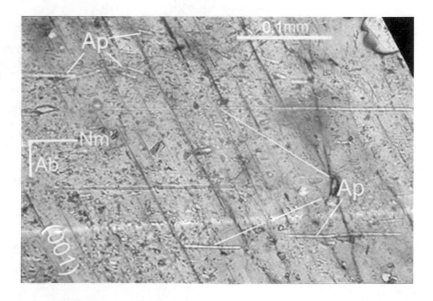

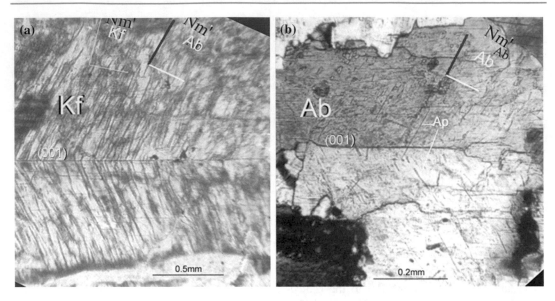

Fig. 3.36 A(−); B(+). Comparison of distribution of perthitic albite lamellae in K-feldspar of Manebach twin (A) with acicular apatite inclusions in albite (An = 4−10) of Manebach twin (B). Shanbei granite, Taishan County, Guangdong Province

Fig. 3.37 (+)Q. Section is roughly //(010). Some of pencil-like albite inclusions in are arranged parallel to the direction of perthitic albite and Nm′. Chengkou granite, Zhuguangshan batholith, Guangdong Province

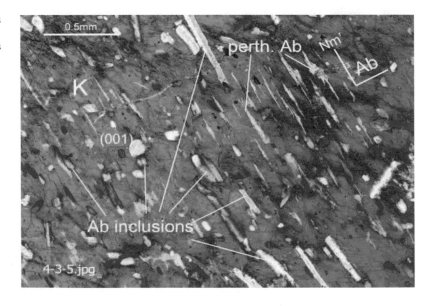

The authors believe that the directional arrangements of long-columnar, pencil like, or acicular crystals enclosed in feldspar (both albite and K-feldspar) are not originated accidently, but due to a special force controlled by the crystallizing growth ability of feldspar itself.

3.3.5 Possible Origin of Perthitic Albite

The phenomenon of oriented inclusion of fine elongate crystals possibly implies that the feldspar minerals growing in granitic melt have a

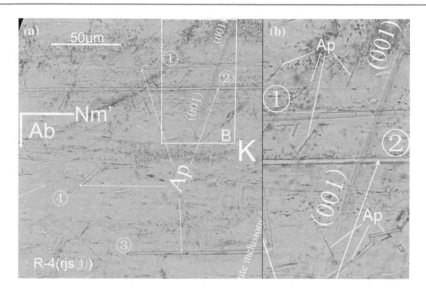

Fig. 3.38 (−). Section ⊥ axis b. B is amplified from part of A. K-feldspar perthite containing inclusions of acicular apatite (Ap). ③ coarse of them ①, ②, and ③ are arranged closely to the Nm′ of albite lamellae. ④ And many fine acicular apatite inclusions are ramdonly distributed. Rössing pegmatoid leucocratic granite, Namibia

Fig. 3.39 Section ⊥ axis b. Back scattered image of K-feldspar perthite containing inclusions of acicular apatite (Ap) (*light white rod*). Some of them are parallel to the albite lamellae (*black rod*). Rössing pegmatoid leucocratic granite, Namibia

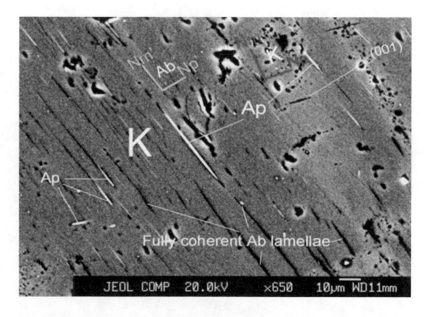

direction of the strongest crystallizing power,[1] which is close to $(\bar{6}01)$, $(\bar{7}01)$, or $(\overline{15}02)$. The power forces the foreign elongated inclusions enclosed parallelly along this direction. Therefore, the perthitic lamellae distributed semi-regularly along the Murchison plane might be formed in the following way:

---

[1]The strongest crystallizing power means the direction of the greatest crystallizing controlling ability which is different from that of the fastest crystallizing speed. The latter is along axes *a* and *c*, while the slowest direction of crystallizing speed is along axis *b* for feldspar minerals. The strongest crystalling power is just coincide with the plane of the minimum value of elastic strain energy, i.e.,

(Footnote 1 continued)
Murchison plane, calculated by Willaime and Brown (1974, 1985).

A few albite crystals could have nucleated dispersively and epitaxially on a crystallizing K-feldspar, i.e., albite component participates in the crystallization of K-feldspar. During further crystallization, the crystallizing albite may settle on either K-feldspar or albite, so does the crystallizing K-feldspar. However, it is generally easier for albite to grow on its own mineral than on K-feldspar. Meanwhile, it is preferentially for K-feldspar to grow on its own mineral than on albite. The direction of parallel arrangement of acicular apatite crystals just implies the strongest crystallization power of the host feldspar, i.e., along the Murchison direction. The gradual and simultaneous crystallization of guest albite with host K-feldspar from melt results in the formation of more or less irregular but roughly parallel arrangements of perthitic lamellae along the Murchison plane ($\overline{1}502$) in K-feldspar.

The authors consider that the vein-like perthitic albite lamellae distributed roughly parallelly and irregularly are more likely formed by simultaneous growth, whereas the fine, dense, parallel, neat perthitic albite lamellae occurring homogeneously and regularly are more possibly formed by solid-state exsolution (Fig. 3.40).

The authors have not studied flame perthite, but temporarily accept that its origin is perhaps related to injection and replacement.

The albite lamellae in metasomatic K-feldspar may occur in two kinds.

(1) Vermicular albite and (2) Semi-parallel or parallel albite.

Figure 3.41 shows that the primary K-feldspars (K_1, K_2, and K_3) are nibbly replaced by the newly replacive K-feldspars (K_4', K_5') in the Huangnitian quartz syenite, Yangjiang County, Guangdong Province. The metasomatic K_4', has two kinds of perthitic albite. The first kind is the coarse vermicular albite several tens μm wide, radiating toward the replaced primary K-feldspar K_1, and the second kind is the thin albite lamellae only <3 μm wide, distributed in dense parallel arrangements. The first kind should be formed simultaneously during replacement, while the second is probably produced by exsolution from the newly formed K-feldspar rather than by replacement simultaneously.

The hetero-oriented K-feldspathization has also been noticed in the Weiya hornblende quartz syenite, Hami County, Xinjiang Autonomous

Fig. 3.40 (+). Thin and dense parallel albite lamellae are probably of exsolution origin, while more irregularly coarse and roughly parallel lamellae are likely formed by simultaneous growth. They are all distributed close to Nm′ of albite

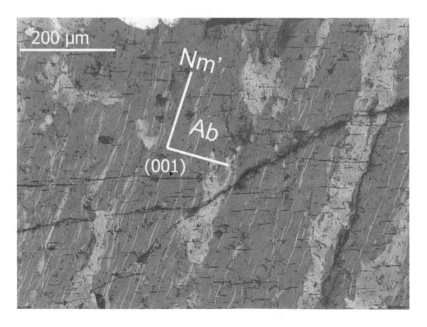

(a)   **(b)**

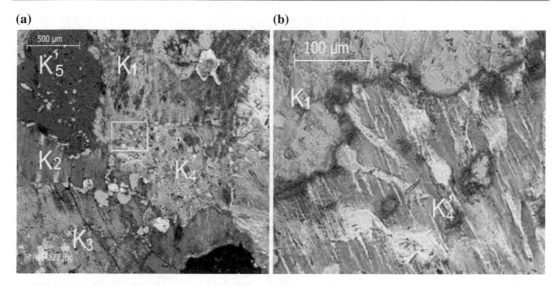

Fig. 3.41 (+)Q. B is enlargement of small frame in A. K4′ is a replacive K-feldspar. The coarse vermicular albite lamellae (tens μm wide) may be formed simultaneously during K4′ replacement, while the thin (<3 μm wide) and dense parallel albite lamellae are probably produced by exsolution. Huangnitian quartz syenite

Fig. 3.42 (+)Q. An old K-feldspar was replaced by a newly formed K-feldspar K′ with scattered relics of perthitic albite Ab. K′ contains perthitic albite lamellae cutting basal cleavage at a great angle, while its left part has vermicular perthitic albite lamellae formed seemingly contemporaneously during metasomatism. Weiya quartz syenite, Hami, Xinjiang

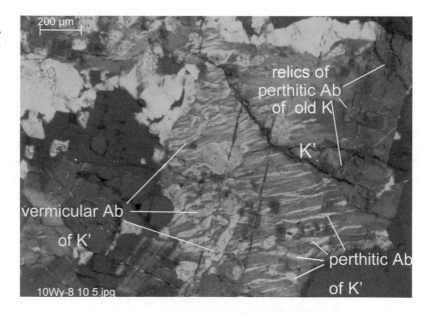

Region. The newly formed K-feldspar K′ has replaced the old K-feldspar K, resulting in the scattered relics of perthitic albite of the latter in K′ (Fig. 3.42).

The left part of K′ in the figure contains many coarse, irregular, vermicular-like albite lamellae, produced probably by simultaneous growth during replacement, while the middle part has semi-parallel albite lamellae which are very similar to the common perthitic lamellae of primary K-feldspar in granite. The semi-parallel albite lamellae may also be formed by either simultaneous growth or unmixing of solid solution.

Fig. 3.43 (+).
Antiperthite. Small
irregular blebs of K_1
present in plagioclase Pl_1
are arranged somewhat
orientedly, showing 60–70°
with (001) in (010) section.
Aqishan granite No. 1,
Shanshan County, Xinjiang
Autonomic Region

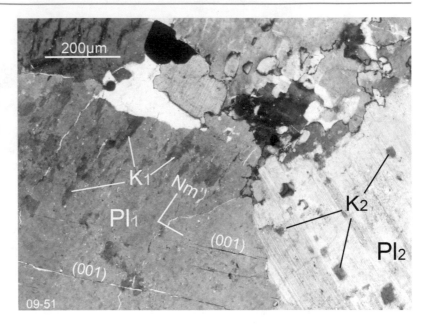

3.3.6   Antiperthite

Antiperthite is intergrowths of guest K-feldspars in host plagioclase, i.e., the inverse of perthite. Antiperthite is commonly seen in granites poor in potassium, but less develops in granites rich in potassium.

Antiperthitic patches in host plagioclase appear usually in form of isolated disseminated and irregular blebs (microcline) with the constant crystallographic orientation strictly as that of the host plagioclase.

Most researchers hold that antiperthite is formed by exsolution process (Hubbard et al. 1965; Vogel 1970).

However, some researchers consider that antiperthite is formed by replacement processes, in which either K-feldspar (microcline) is co-orientedly replaced by albite (plagioclase) (Plümper and Putnis 2009) or plagioclase is co-orientedly replaced by K-feldspar (Dong and He 1987; Collins 2002).

It is hardly possible that K-feldspar could be co-orientedly replaced by plagioclase and plagioclase be co-orientedly replaced by K-feldspar in authors' consideration.

The authors have not deeply studied antiperthite, but they noticed that it occurs mostly as flattened rod-shaped blebs (Fig. 3.43) rather than vein-like albite lamellae in the host K-feldspar of perthite. However, the antiperthitic K-feldspar blebs at the upper part of Pl_1 in Fig. 3.43 have elongated shapes, the direction of which is crossing at 60–70° with cleavage (001) and approximately parallels Nm', i.e., the projection of Nm of plagioclase on (010).

Therefore, the authors suggest an origin of simultaneous growth for antiperthitic blebs of coarse and irregular form with the host plagioclase, in addition to exsolution origin.

3.4   K-Feldspar Phenocryst or Porphyroblast

Porphyritic texture is often present in granite. Short coloum shaped phenocryst is always randomly distributed, while long plate-like one may generally arranged orientedly.

In granophyres or microfine grain porphyritic granites few inclusions is contained in the euhedral K-feldspar phenocryst, while in medium-coarse grain porphyritic granites several inclusions of plagioclase and biotite may be contained in the phenocryst with less euhedral form, and the granite might be named as "porphyroid" granite, as if the origin of the granite is uncertained and the granite may be formed by metasomatism. Especially, when granite demonstrates gneissic or augen structure, the granite is often nominated as migmaitic granite or augen gneissic granite. The augen is mainly composed of K-feldspar, then the augen K-feldspar might be treated as porphyroblast.

One key point regarding the origin of granitic rocks is the origin and crystallization of feldspars, in particular K-feldspar, because it is one of the most important mineral phases in granitic rocks. The controversy about the origin of granite actually is a controversy about the origin and formation of K-feldspars. Therefore, it is necessary to correctly distinguish the difference between K-feldspar phenocryst and porphyroblast for determination whether the rock is formed by metamorphic or magmatic processes.

Both phenocryst and porphyroblast are relatively large crystals compared to the sizes of grains in the ground mass. A phenocryst is crystallized from a magmatic melt, while a porphyroblast is formed by metasomatism during metamorphism. Both may or may not contain mineral inclusions of the ground mass.

3.4.1 Theory of Porphyroblast Origin

There are two major hypotheses: metamorphic autocathasis and metasomatic genesis.

3.4.1.1 Metamorphic Autocathasis

Autocathasis described by Harker (1950) in his classical book "Metamorphism" is as follows: "Growing crystals endeavor to clear themselves by expelling foreign inclusions of any kind, but their power to do so depends upon their inherent force of crystallization…An instructive case is that in which a growing crystal has been able to

brush aside foreign material, but not completely to eject it, the result being that in the grains of inclusions remain caught in the crystal along certain directions in which the force was least effective".

Augustithis (1973) emphasized "Often a crystalloblast tends to clear itself from inclusions by "autocatharsis" (self cleaning process) either by pushing them at its periphery or confining them along certain crystallographic directions within the crystalloblasts. In cases, the inclusion of groundmass components (e.g., myrmekitised plagioclases and corroded biotites) in K-feldspar crystalloblasts follows exactly the (001) (010) (110) face". The certain crystallographic directions are those in which the force was least effective, as described by Harker (1950) in his book "Metamorphism".

These points of view are doubtful. The surviving relicts should be preserved in situ without changing their own primary crystallographic orientations, because the metasomatic process is carried out when a rock is in a solid-state. Even if plastic deformation occurs during dynamic metamorphism, the unreplaced residue might be deviated and/or rotated slightly. Nevertheless, it is still impossible to push the unreplaced relicts aside and to rearrange them along the least effective crystallographic directions.

3.4.1.2 Metasomatic Genesis
(1) **Hippertt hypothesis**
Hippertt (1987) studied the metasomatic replacement of plagioclase by microcline and the development of a microcline porphyroblast in augen gneiss, Niteroi, Brazil, and showed a serial drawing which is cited in "Rock-forming minerals" (Deer et al. 2001 edition, page 585, Fig. 417). He postulated the original rock and sequence of metasomatic process as shown in the drawing (see Fig. 3.44) as follows:
(a) The original metamorphic rock is composed mainly of plagioclase crystals;
(b) During the beginning of microclinization the small plagioclase crystals recrystallized around microcline;

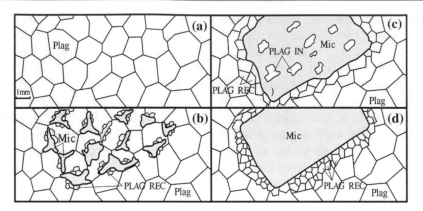

Fig. 3.44 Serial drawings showing the metasomatic replacement of plagioclase by microcline and the development of microcline porphyroblasts in augen gneiss, Niteroi, Brazil (after Hippertt 1987), transferred from "Rock-Forming Minerals" (Deer et al. 2001, p. 585). *Mic*—microcline, *Plag*—plagioclase, *In*—inclusion, *Rec*—recrystallized

(c) Some adjacent crystals of microcline have merged into one porphyroblast enclosing some plagioclase inclusions arranged with long dimensions roughly parallel to lattice planes of the microcline;

(d) The porphyroblast is transformed into an idiomorphic shape and able to replace thoroughly the enclosed plagioclase inclusions.

The major problems of the above description of the metasomatic processes are as follows:

Why are most plagioclase grains completely replaced by microcline while the minor plagioclase grains may survive from microclinization and remain unchanged, and why are the plagioclase grains arranged roughly parallel to lattice planes of the microcline?

How can several small anisotropic microcline crystalloblasts merge to form a large one with the same or similar size and form?

Why are the anisotropic elongated megacrysts arranged roughly parallelly?

(2) **Collins hypothesis**

Collins (2002) investigated K-feldspar megacrysts in Papoose Flat pluton in the Inyo Mountains of California, USA. He suggested that orthoclase megacrysts are formed by K-Si-metasomatism after brittle deformation caused by local cataclasis which produced avenues in rock so that K- and Si-bearing fluids could be introduced. At first, growing on the preexisting K-feldspar the new K-feldspar began to replace plagioclase, quartz, and biotite grains. Most of the strongly broken or unoriented plagioclase crystals and the groundmass minerals (quartz and biotite) would be preferentially replaced, whereas those plagioclase crystals whose lattice orientation coincides with that of the replacing orthoclase would be mostly preserved, unchanged. Continued deformation allowed fluids to move through the system, causing K-feldspathization. Following the formation of the first orthoclase crystals, the repeated cataclasis and K-feldspar metasomatism helped the growth of the euhedral megacrysts of orthoclase, forming zonal alignment of plagioclase crystals parallel to Ba-K growth zones within megacrysts (Figs. 3.45 and 3.46).

There are several doubtful points:

First, microcline megacryst contains only quartz <3 % and biotite 1–2 %, much less than the contents in primary rock, which are 10–20 % and ∼6 %, respectively. The question is whether a lot of quartz and biotite are really replaced by microclinization? Microscopic observation proves neither quartz nor biotite is irrefutably replaced by K-feldspathization either in rocks or in K-feldspar megacryst.

Second, according to the rule of hetero-oriented replacement, K-feldspathization is indeed unable to replace the co-oriented plagioclases. However,

Fig. 3.45 (+). Orthoclase megacryst, showing well developed concentric banding of tiny parallel plagioclase inclusions. Their crystallographical orientations are different from the host K. Papoose Flat pluton, California, USA. From Collins (2002). (With kind permission from Collins)

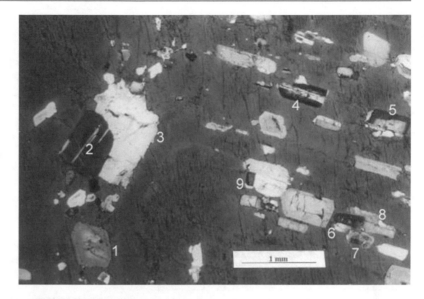

Fig. 3.46 Orthoclase crystal from pluton cut through center parallel to (010). *Left* Portion unstained, *right* portion *stained red* for Ba. Layers oscillatory zoned with regard to Ba and K. Hour glass shapes show preferential concentrations of minerals included in face sectors. Papoose Flat Pluton, California (Dickson 1966)

nearly all plagioclase inclusions oscillatorily aligned in megacryst are not co-orientedly enclosed in the host megacryst, since they have different interference color from that of the surrounding perthitic albite lamellae (Fig. 3.45) It would be clearly noticed with quartz plate inserted and can be further proved if either myrmekite or albite rims appear on the border of the plagioclase inclusions. In fact, these inclusions are enclosed just with their long dimensions roughly parallel (aligned) to the growth (crystallizing) plane of the host megacryst. It is obviously doubtful that

microclinization could totally replace the plagioclase inclusions randomly distributed but did not replace those concentrically aligned, although they all had the different orientation from that of the host megacryst.

Third, were all the unoriented plagioclase crystals really replaced by microclinization completely, without any traces at all?

Surely, the replacement of enclosed plagioclase crystals by K-feldspar could certainly occasionally be observed inside a megacryst of K-feldspar. Many petrographers insist on treating

this phenomenon as an important criterion to
determine that the whole K-feldspar has a porphyroblast origin.

Nevertheless, the degree of replacement of the
plagioclase by K-feldspar is more or less comparable within the thin sections. There is no
gradual or gradational transition from plagioclase
being slightly replaced to that being thoroughly
replaced in thin sections. In fact, what we see is
that plagioclase is replaced either partly or
"completely" without gradual transition. In
addition, plagioclase is unable to be co-orientedly
replaced by K-feldspar. Hence, there is short of
evidence for the complete replacement of all
hetero-oriented plagioclase grains.

Therefore, the pure part of K-feldspar megacryst, i.e., the part without relict of plagioclase
inclusions can hardly be judged as porphyroblast
formed by K-feldspathization which has thoroughly replaced plagioclase grains.

3.4.2   Theory of Magmatic Origin

Plagioclase crystal in granite and granodiorite is
more idiomorphic than K-feldspar and may be
enclosed or semi-enclosed in K-feldspar. Several
laboratory experiments (Luth 1976; Fenn 1977;
Burnham et al. 1971, 1974) show that plagioclase begins to crystallize earlier than K-feldspar
in melt during decreasing temperature.

Swanson (1977)[2] carried out an artificial
experiment of granitic and granodioritic melt to
investigate the nucleation density and growth
rate of major minerals in synthetic granitic and
granodioritic melt. The results show that at small
undercoolings (ΔT (T(**Liquidus**) − T(**growth**))
<100 °C) in granitic melt (Fig. 3.47), the growth
rate of alkali feldspar is much faster, say several

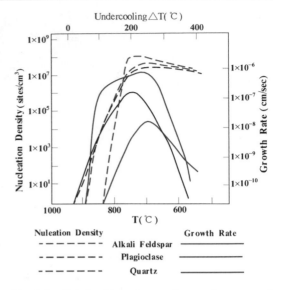

Fig. 3.47 Relationship of crystal growth rates and
nucleation densities curves with undercooling in a granitic
melt with 3.5 wt% H_2O at 2.5×10^8 Pa. Crystallizing at
low degree of undercooling ΔT(T(liquidus)-T(growth)),
alkai-feldspar has much faster growth rate compared with
plagioclase, while their nucleation densities are in the
same grade. (After Swanson 1977)

to tens times, than that of plagioclase, while their
nucleation densities are approximately in the
same grade, causing the formation of alkali-
feldspar phenocryst. Therefore, it is possible that
inclusions of small plagioclase crystals may be
contained in alkali-feldspar phenocryst.

Vernon (1986, 2002) considered that megacryst K-feldspar in deformed granites is phenocryst relic instead of porphyroblast. He
proposed the following major criteria of identification for K-feldspar phenocryst: (1) euhedral;
(2) simple twin; (3) zonal inclusion of idiomorphic biotite and plagioclase; and (4) oscillatory
zonal change of composition (especially Ba)
from inside to outside.

Wang et al. (2002) studied microcline megacrysts in Fogang granitic complex, South China
and believed that the megacrysts are formed by
magmatic rather than metasomatic origin.

The authors agree with the above authors's
idea and stress that the K-feldspar megacryst
cannot be considered to be basically formed by
metasomatic processes even in case of actual
presence of the hetero-oriented replacement

[2]Analysis of crystal growth as a function of temperature
for synthetic granite and granodiorite compositions in the
system $KAlSi_3O_8$–$NaAlSi_3O_8$–$CaAl_2Si_2O_8$–SiO_2 under
H_2O-saturated and undersaturated conditions yields quan-
titative data on the growth kinetics of plagioclase,
alkali-feldspar, and quartz in these systems at ≤ 8 kbar
and 400–900 °C. Measured growth rates vary from
3×10^{-6} cm/sec (2.6 mm/day) to 1×10^{-10} cm/sec
(about 0.03 mm/yr), while nucleation density (nucleation
sites/unit volume) varies from 0 to over 1×10^8 sites/cm^3.

phenomena of plagioclase by K-feldspar in the megacryst. It is emphasized to carefully observe and compare the development of perthitic lamellae in the K-feldspar. The perthitic lamellae in the K-feldspar (K') area closely surrounding the relics of replaced plagioclase inclusions are generally less developed than that in the rest area of K-feldspar megacryst (Figs. 1.15 and 1.16).

According to the rule of hetero-oriented replacement, leaning on the preexisting K-feldspar and taking its orientation, the newly formed K-feldspar replaces the hetero-oriented plagioclase. There is no clear boundary between the preexisting K-feldspar and the newly formed replacive K-feldspar.

Generally, it is possible to distinguish the replacive K-feldspar from the primary K-feldspar by the development degree of the perthitic lamellae between them.

A wrong impression that the whole K-feldspar megacryst is metasomatic will be easily made if the primary K-feldspar contains nearly no perthitic albite lamellae (in intermediate rocks with more Ca, Fe, and Mg and less SiO_2), because there is no divergence between the primary and the replacive K-feldspars under microscope.

There is no gradational transition from slightly replaced plagioclase to thoroughly replaced one in thin sections. In other words, what we see is that either plagioclase is slightly or locally replaced by K-feldspar (with small relics of plagioclase), or K-feldspar is pure and clear without any relics of plagioclase. Therefore, the authors insist that the pure and clear part of K-feldspar represents the preexisting K-feldspar rather than the product of complete replacement of plagioclase.

The authors consider that K-feldspar megacryst is still of primary origin, although plagioclase inclusions have been locally superimposed by the hetero-oriented K-feldspathization inside the K-feldspar megacryst.

The authors would point out here that "K feldspar megacryst is phenocryst or residual phenocryst rather than porphyroblast" does not deny the real existence of various porphyroblast in rocks. For example, andalusite presenting in carboniferous hornfels after contact metamorphism,

garnet, stauvrolite, etc., appearing in regional metamorphic rocks are formed by metasomatism.

3.5 Study of History of Metasomatic Processes

The texture of a certain rock may become complicated due to the superimposition of the newly formed minerals after several metasomatic processes. The traces and relationship of mineral replacement, however, may help to discover the successive history of replacement during the evolution of rocks which is one of geological problems needed to be solved.

The successive sequence of mineral replacement can reasonably be deduced according to the rule of mineral replacement and the contact relationship among the replacive and replaced minerals. Some practical examples are as follows:

3.5.1 Hetero-Oriented Albitization Occurs Earlier Than Quartzification

Both hetero-oriented albitization and quartzification are observed in leucocratic alkali-granite (Beihuan granite, Yangjiang County, Guangdong Province). The granite consists of albite 40 %, quartz 37 %, alkali-feldspar 19 %, and protolithionite 3 %. K-feldspar was intensively replaced by either albite or quartz. The growth width of the replacive quartz (>0.5 mm) is greater than that of the replacive albite (<0.5 mm). The relics of the swapped albite rows $Ab_1'Ab_2'$ are fortunately noticed by inserting quartz plate under crossed polars (Fig. 3.48).

The contact line between Ab_1' and Ab_2' should be the previous border of the preexisting K-feldspar K_1 and K_2. The replacive quartz Q_1' Q_2' evidently damaged the connection of Ab_1' Ab_2' with the substrate K_1 and K_2. Therefore, hetero-oriented albitization should have occurred prior to quartzification. Otherwise the swapped albite rows could not be formed on the background of replacive quartz.

Fig. 3.48 (+)Q. Swapped albite $Ab_1'Ab_2'$ rows occurred first, followed by quartzification $Q_1'Q_2'$. Dotted line shows rough sketch of K_1 Ab_1 and K_2 Ab_2. Shanbei leucogranite, Taishan County, Guangdong Province

3.5.2   Swapped Albite Rows Are Formed First, Followed by Beryllization

Small pegmatite miarolitic cavities are distributed in the apical facies of a leucogranite intrusion (Shanbei granite, Taishan County, Guangdong Province). These cavities are filled with the anhedral beryl (Be) and are surrounded by euhedral K-feldspar and plagioclase. The irregular beryl crystals (Be') that enclose perthite albite lamellae are scattered in the K-feldspar. The anhedral beryl (Be) outside the euhedral K-feldspar is primary (Figs. 1.54, 1.55 and 3.49), while the irregularly distributed beryl (Be') (inside the K-feldspar) with the same orientation as that of the anhedral beryl (Be) outside the euhedral K-feldspar is metasomatic (Fig. 3.50).

The swapped rows of albite (Ab_1' and Ab_2' in the middle of Fig. 3.49, Ab_1' and Ab_3' in the middle of Fig. 3.50) are situated and still remain at the grain boundary between the residual

Fig. 3.49 A(+); B(+)Q. The anhedral beryl (Be) outside the euhedral K-feldspar is primary, while the irregularly growing beryl (Be') inside the K-feldspar is of replacement. Shanbei leucogranite, Taishan County, Guangdong Province

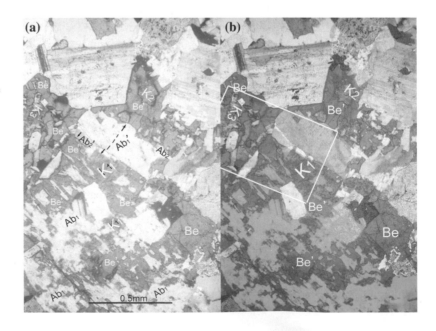

Fig. 3.50 (+)Q.
Beryllization Be′ was
formed later than the
swapped rows of albite
Ab$_1$′Ab$_2$′ as the latter are
surounded by the former

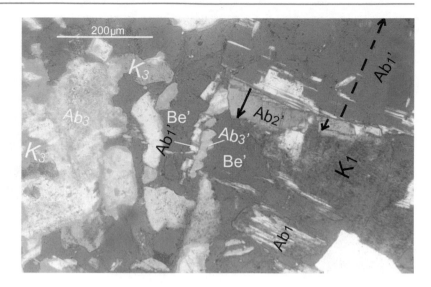

K-feldspars K$_1$ and K$_2$ or K$_1$ and K$_3$, which are partly replaced by Be′. The tiny swapped rows of albite Ab$_1$′ and Ab$_3$′ must have formed earlier than beryllization, because they are surrounded by metasomatic beryl Be′.

3.5.3   Triple Nibble Replacement

Formed during the third period of the Indosinian cycle, in the Zhuguangshan batholith, Guangdong Province, the earlier granitic intrusion Huang-shatang medium-fine grained porphyritic two-mica granite experinced three times of hetero-oriented replacement (Figs. 3.51 and 3.52). First, double rows of swapped albite (Ab$_1$′ Ab$_2$′) (0.3–0.4 mm thick) sericitized later were formed by the albitization replacing the K-feldspars (K$_1$, K$_2$). The second replacement was the K-feldspathization (K$_1$″, K$_2$″) replacing the swapped albite. The K$_1$″ and K$_2$″ with 0.2–0.3 mm in thickness have less perthitic albite

Fig. 3.51 (+).
Complicated contact at the
boundary between two
K-feldspars (K$_1$, K$_2$)

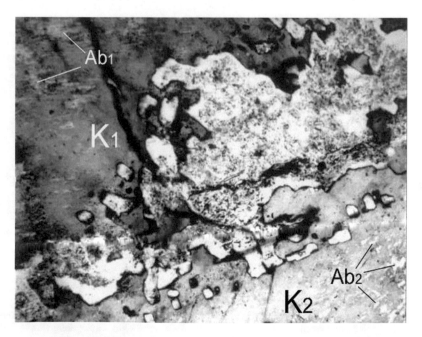

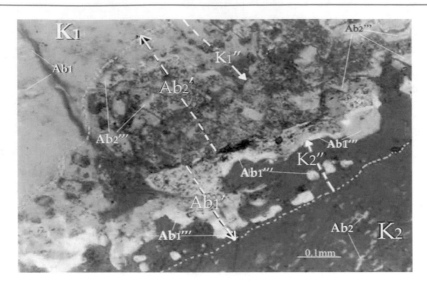

Fig. 3.52 (+)Q. Triple replacement. Swapped albite rows (Ab$_1'$, Ab$_2'$) took place first, followed by second K-feldspathization (K$_1''$, K$_2''$) that replaced the swapped albite. Finally, the third slight albitization (Ab$_1'''$, Ab$_2'''$) replaced the K-feldspar (K$_1''$, K$_2''$). Huangshatang granite, Zhuguang batholith, Guangdong Province

than that of the primary K-feldspars K$_1$ and K$_2$. The third replacement process was the slight albitization producing albite clear rims (0.1 mm thick) (Ab$_1'''$, Ab$_2'''$) which replaced the K-feldspar (K$_1''$, K$_2''$). The clear rims appeared only at contacts between the relics of swapped albite rows and the K-feldspar with different orientation. No clear rim emerged at contacts between two sericitized albite rows.

There is another example wherein several small albite crystals appeared at the contact of two K-feldspars (Fig. 3.53a).

These scattered albite crystals are obviously divided into two groups when the quartz plate is inserted under crossed polars (Fig. 3.53b). Each group is located in an individual K-feldspar and has similar or nearly the same crystallographic orientation as that of the perthitic albite of the

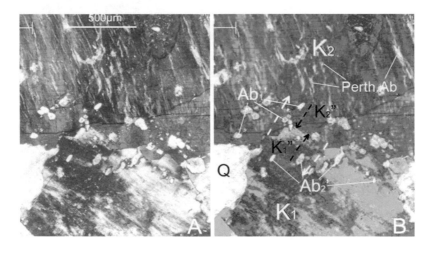

Fig. 3.53 A(+); B(+)Q. At the border of two K-feldspars occur mess tiny albite crystals, which are divided into two groups, when quartz plate is inserted. Each group holds the similar orientation as that of perthitic albite of opposite K-feldspar, resulting from replacement processes in twice. Firstly, swapped rows of Ab$_1'$ and Ab$_2'$. Secondly, reverse K-feldspathization K$_1''$ and K$_2''$ (also about 0.3 mm thick). Chengkou granite, Zhuguang batholith, Guangdong Province

Fig. 3.54 (+)Q. $K_1'' K_2''$ lacks perthitic albite lamellae compared with K_1 K_2. The zigzag boundary of K_1'' with K_2'' is not the original one between K_1 and K_2

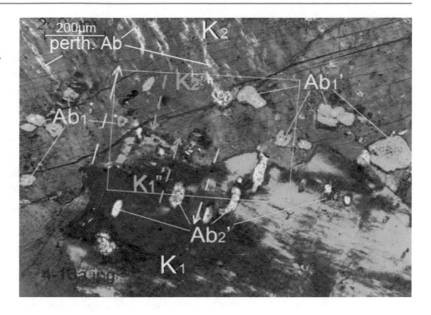

opposite K-feldspar. It can be asserted that two major replacement processes occurred here. First, the hetero-oriented albitization happened, forming the swapped rows of Ab_1' and Ab_2' (0.2 mm thick). Then, the hetero-oriented K-feldspathization took place, forming K_1'' and K_2''. The thickness of the newly formed K-feldspar is also about 0.2 mm, resulting in the disappearance of most parts of the swapped albite rows and the presence of their isolated relicts. With careful observation, in the area of K_2'' some tiny water-drop quartz crystallites Q (5–20 μm in diameter) as relicts of myrmekitic quartz of swapped albite rows can be noticed (Fig. 3.55), which are absent in ordinary primary K-feldspar. Having completely replaced Ab_1' Ab_2', the replacive K-feldspars $K_1''K_2''$ met together and replaced each other, causing the

Fig. 3.55 (+)Q. Enlargement of Fig. 3.54. In $K_1'' K_2''$ there are tiny water-drop quartz crystallites Q, which are relicts of myrmekitic quartz contained in swapped albite rows

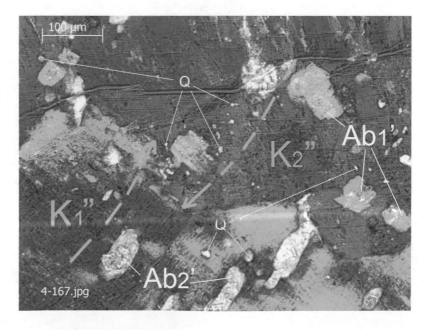

transformation of the initial border between the two original K-feldspars.

There is still tiny albite rim (third time replacement) (only 10–20 μm thick) formed on the border of relics $Ab_1'Ab_2'$ with hetero-oriented K-feldspar crystals $K_1''K_2''$.

The biotite was co-orientedly replaced by chlorite. However, it is difficult to judge the successive sequence of the two kinds of meta-somatism (hetero-oriented muscovitization and co-oriented chloritization), because there is no direct relationship between them.

3.5.4 Hetero-Oriented Muscovitization and Co-oriented Chloritization of Biotite

A complicated texture appears at the contact of a primary euhedral biotite (transformed into chlorite later) with a muscovite (Fig. 3.56).

The zigzag but roughly straight contact of biotite with muscovite indicates that the biotite is idiomorphic while the muscovite is xenomorphic. The muscovite is evidently divided into two parts, although they have the same interference color and the same orientation, and there is no clear boundary line between them. The outside major part Ms contacted mainly with quartz should be of primary and the inside minor part Ms' penetrating irregularly into biotite (chlorite) must be of metasomatic.

3.5.5 Hetero-Oriented Albitization, Protolithionitization, Muscovitization and Quartzification

In a leucogranite Beihuan albitic granite rich in alkaline and silica, Yangjiang County, Guang-dong Province, the alkali-feldspar is composed of intergrowth crystals of K-feldspar with extensively distributed albite (Fig. 3.57).

Hetero-oriented albitization, protolithionitiza-tion, muscovitization, and quartzification are widely developed in the granite and the K-feldspar phase is the major commonly replaced mineral for the above-mentioned meta-somatic processes.

Figure 3.58 shows the location of three alkali feldspars K_1, K_2, and K_3 with protolithionite and quartz.

Fig. 3.56 (+). Euhedral biotite (chloritized) with complicated muscovite. Muscovite can be divided into primary Ms and replacive Ms'

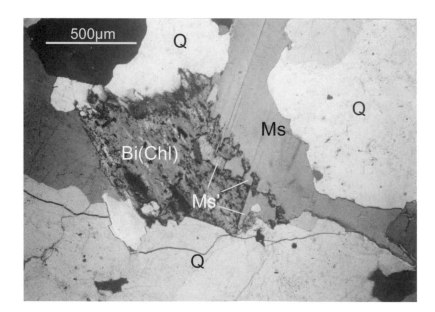

Fig. 3.57 (+)Q. Alkali feldspar K_1 K_2 consists of K-feldspar with abundant perthitic albite. Beihuan granite

Fig. 3.58 (+). Protolithionitization Bi′ and quartzification Q′ have deeply replaced alkali feldspar, leaving over relics of perthitic albite. Beihuan leucogranite, Yangjiang County, Guangdong Province

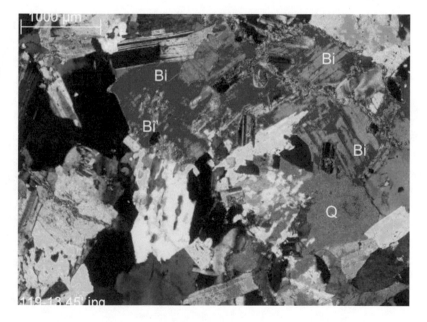

Three albitic skeletons of alkali feldspar grains remained after the K-feldspar phase had been thoroughly replaced (Fig. 3.59).

The swapped albite rows $Ab_1′Ab_2′$ and perthitic albite Ab_1 are separated by metasomatic protolithionite Bi′ and quartz Q′, indicating that Bi′ and Q′ are formed later than the swapped albite $Ab_1′Ab_2′$.

Figure 3.60 shows three grains of protolithionite $Bi_1′$, $Bi_2′$, Bi_3. With quartz plate inserted (Fig. 3.61), it is clearly seen that the relics of perthitic albite are contained in $Bi_1′$ and $Bi_2′$, so they should be of metasomatic, i.e., the products of protolithionitization. The Bi_3, located outside the area of relics and bordered with quartz and idiomorphic plagioclase,

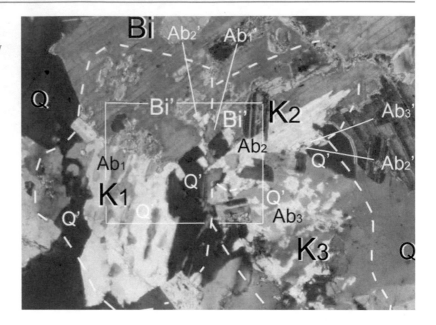

Fig. 3.59 (+)Q. Three primary K-feldspar were placed by Bi′ and Q′. They were formed later than swapped rows of albite

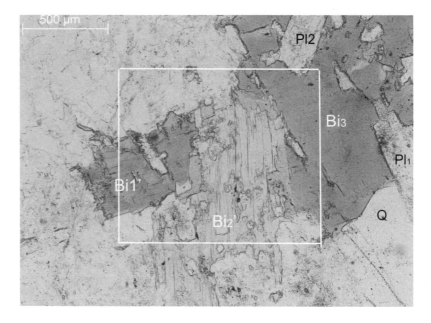

Fig. 3.60 (−). Weak pleochroism of protolithionite $Bi_1′$ $Bi_2′Bi_3$ in Beihuan leucogranite

may not be metasomatic, but probably primary in origin.

Figure 3.62 is an enlargement of inner frame of Fig. 3.60 which shows that the swapped muscovite grains $Ms_1″$ $Ms_2″$ appear at contacts of $Bi_1′$ with $Bi_2′$ and the replacive muscovite $Ms_3″$ occurs unidirectionally toward $Bi_2′$.

Therefore, muscovitization took place later than protolithinitization.

K-feldspar was replaced by both protolithionite and quartz, and there is no relict relationship among them. So, it is hard to judge which one of them happened earlier. However, on the basis of the following two facts: ① the

Fig. 3.61 (+)Q. Bi_1', Bi_2' and Q' containing relicts of perthitic albite must be of metasomatic, while pure Bi_3 may probably be of primary

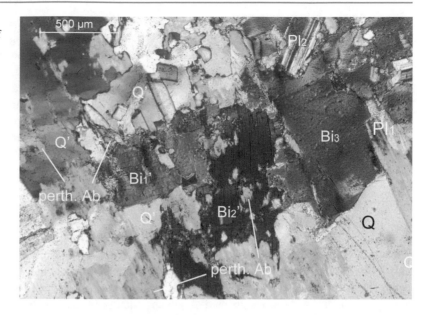

Fig. 3.62 (−). Magnified from Fig. 3.60. The swapped muscovite Ms_1'' Ms_2'' occur at the contact between Bi_1' and Bi_2'. Ms_3'' unilaterally replaces Bi_2' at the border of Bi_3 with Bi_2'

contact area of K-feldspar with primary quartz much more than that with primary protolithionite; ② the replacive growth rate and amount of replacive quartz faster and greater than that of protolithionite (Figs. 1.46 and 1.47), the authors guess that the protolithionitization might possibly occur earlier than quartzification, otherwise, it would be no place for the protolithionitization to occur at all.

3.6   Successive Sequence of Multi-metasomatic Processes in Alkali Metasomatites

There were several replacement processes, such as co-oriented albitization of K-feldspar, calcitization, hetero-oriented albitization, quartzification, etc., that occurred in the sodium metasomatites in Jiling medium-coarse grain porphyritic granite, Gansu Province, North China.

According to replacement texture and mutual contact relationship the successive sequence of these replacive processes can be rationally estimated as follows:

3.6.1 Co-oriented Albitization Later Than Earliest Hetero-Oriented Albitization

Hetero-oriented albitization of K-feldspar occurred at the deuteric stage soon after intrusion, while the co-oriented albitization of K-feldspar took place in the sodium metasomatites which were locally distributed in the intrusion body.

Figures 3.63, 3.64 and 3.65 indicate that swapped albite rows were present at the contact between the resembling K-feldspars which were thoroughly transformed into albite by co-oriented albitization.

The swapped albite rows would not occur if the two K-feldspar crystals had been transformed into chessboard albite. Therefore, the hetero-oriented metasomatic albite must have been produced prior to the co-orientation

albitization of K-feldspar. This relationship is in accordance with geological observation, because swapped albite can be seen everywhere in the whole rock body and must have been formed during a postmagmatic process soon after intrusion, while the transformation of K-feldspar into chessboard albite occurred only in local regions (i.e., sodium metasomatite bodies) after all stages of the granitic magmatic intrusions had ceased.

3.6.2 Calcitization Replacing Quartz and K-Feldspar

Calcite might replace quartz and K-feldspar, and thus become a major rock-forming mineral in sodium metasomatites. The disappearance of quartz in granite due to intense replacement of quartz by calcite is one obvious feature of sodium metasomatites, where K-feldspar may be completely replaced or transformed by co-oriented albitization.

Then, among the two metasomatic processes which one occurred first, the calcitization of quartz and K-feldspar or the co-oriented albitization of K-feldspar?

Microscopic observation indicates that calcitization might replace quartz and K-feldspar, but did not easily replace small albite grains.

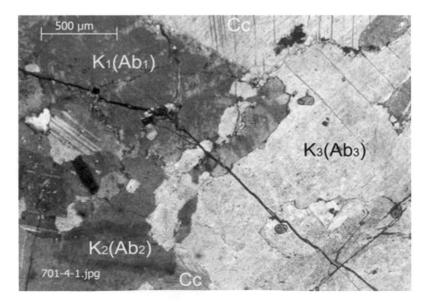

Fig. 3.63 (+). Swapped albite rows occur at contact between ghost K-feldspars (co-orientedly albitized)

Fig. 3.64 (+)Q. Swapped albite rows must have occurred earlier than the co-oriented albitization of K-feldspar. Sodium metasomatite in Jiling granite

Fig. 3.65 A(+); B(+)Q. Swapped albite rows must have occurred earlier than the co-oriented albitization of K-feldspar. Sodium metasomatite in Jiling granite

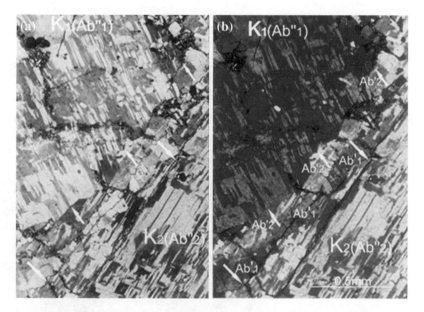

Figure 3.66 shows that calcite did not replace the whole plagioclase as well as chlorite (former biotite).

Therefore, the scope of messy small albite crystals with the same orientation in calcite background must be the relicts derived from K-feldspar.

Where did these albite relicts come from? Did these albite relicts come wholly from perthitic albite or partly from co-oriented albitization?

If these albite relicts were all derived from perthitic albite lamellae only, and no albite relics from co-oriented albitized albite, it means that the K-feldspar was replaced by calcite prior to the co-oriented albitization of K-feldspar. In other words, calcitization should occur earlier than the co-orientation albitization of K-feldspar.

If much of the albite relicts basically belonged to albite grains formed by co-oriented albitization, besides perthitic albite lamellae, there

Fig. 3.66 (+)Q. Calcite
Cc has strongly replaced
K-feldspar with many
relicts of albite remained,
but not replaced block
plagioclase and chlorite.
Sodium metasomatite in
Jiling granite

would be a contrary deduction that the co-oriented albitization of K-feldspar must be prior to the calcitization.

The co-oriented albitized albite is characterized by its distribution pattern, beginning with group dotted, to slice form, then to lump form, that is quite different from the perthitic albite lamellae (mainly along Murchison plane $(\overline{6}01)$,

$(\overline{7}01)$ or $(\overline{1}502)$ (Figs. 1.68, 1.69, 1.70, 1.71, 1.72 and 1.76).

In Fig. 3.67, the albite relicts extending along Ng' (in section close ⊥(010)) belong possibly to perthitic albite. Such relicts, however, are rarely observed.

However, the majority of relicts are prismatic (Fig. 3.68) or lumpy. Especially the elongation

Fig. 3.67 (+)Q.
K-feldspar K is strongly
replaced by calcite Cc.
Since the elongation of the
relicts of albite is close to
Ng', they may be of
perthitic albite lamellae

Fig. 3.68 (+)Q. The elongation of residual albite grains along Np' indicates that they are relicts of co-oriented albitized albite. Sodium metasomatite in Jiling granite

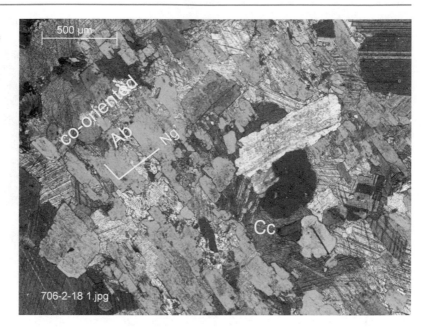

of the prismmatic relicts is negative (close to Np') in section ⊥(010) (Fig. 3.68), indicating that they are extended along (010) direction. It shows that these relics are derived from co-oriented albitized albite rather than perthitic albite.

If K-feldspar was completely co-orientedly replaced by albite and transformed into a whole albite crystal, the latter could not be replaced by calcite, as shown in the upper and lower parts of

Fig. 3.63, in which calcite, having completely replaced quartz, was unable to replace the co-orientedly albitized albite from K-feldspar. Therefore, it can be judged that co-oriented albitization of K-feldspar must occur earlier than calcitization.

Figure 3.69 shows two groups of albite relicts derived from two replaced K-feldspar crystals. At the contact between the two groups there are

Fig. 3.69 (+)Q. Swapped rows of albite at boundary of the two piles of albite grains. The order of metasomatic processes is judged as: 1 swapped rows of albite between two K-feldspars; 2 the latter partly co-orientedly albitized (with negative elongation in section //axis *b*); 3 the residual K completely replaced by calcite. Sodium metasomatite in Jiling granite

Fig. 3.70 (+). K-feldspar was strongly co-orientedly albitized while the primary quartz Q was preserved

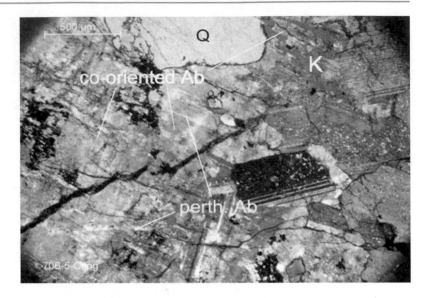

swapped albite rows $Ab_1'Ab_2'$. The section herein is roughly perpendicular to (010) on the basis of the albite twin. The fact that the elongation of albite grains is parallel to the albite twin plane (010) shows they were derived from co-oriented albitized albite rather than perthitic albite.

It may be inferred that the hetero-oriented albitization occurred evidently earliest, forming swapped rows of albite at the boundary between two K-feldspar grains. After a long time, the co-oriented albitization of K-feldspars took place, but not completely, preserving residual K-feldspar, which could be partly, even totally replaced by calcite later.

Figures 3.70, 3.71 indicate that K-feldspar was intensely co-orientedly albitized to chessboard albite, while the primary quartz was still completely preserved without any trace of replacement by calcite. Therefore, no calcitization occurred herein till the completion of co-oriented albitization of K-feldspar.

Fig. 3.71 (+). K-feldspar was thoroughly co-orientedly albitized while primary quartz Q survived

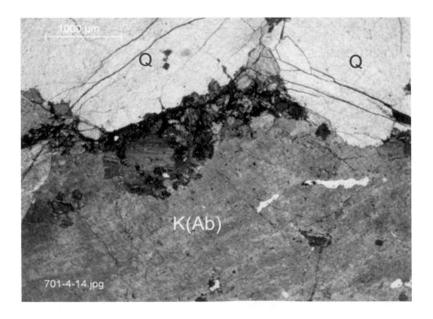

In the light of the existence of a transitional zone of the co-oriented albitization from mild (group dotted form) → to moderate (slice form) → to extensive (lump form), as well as of the replacement of the residual K-feldspar by calcite, the distinguishable difference in distribution pattern between the different degrees of co-oriented albitized albite and perthitic albite helps to get an incontestable conclusion that the co-oriented albitization must be prior to the calcitization. Otherwise, the successive order of the two replacement processes (co-oriented albitization and calcitization) could not be ascertained, if the transitional zone did not exist as in Fig. 3.63.

3.6.3 Late Stage Clear Rim Albitization Replacing Calcite

After calcitization, the late stage hetero-oriented albitization happened at the contact of calcite with various albite crystals (Fig. 3.72), forming a clear albite rim with thickness up to 200 μm, which is much more greater than that of the early stage clear albite rim toward K-feldspar.

The late stage albite clear rim is commonly present around the border of albitic relicts with calcite, making the relicts thicker and flatter (Fig. 3.73).

No metasomatic relics of calcite are found in the late stage albite clear rim perhaps because calcite can be easily replaced by albite.

3.6.4 Albite Microvein and Vein Selvage Albitization Replacing Calcite

Tiny albite microveins (5–100 μm thick) in co-orientedly albitized albite (transformed from K-feldspar and plagioclase) and calcite occurred within microbroken places of the Jiling alkali metasomatites (Fig. 3.74).

Albite microvein is brown in color due to micro particles of hematite. From the bilateral selvage of microvein in calcite, pure albite grains grew toward calcite, taking the same orientation of albite in microvein, with thickness up to 150 μm (Figs. 3.75, 3.76). The extent of universal growths from the selvage of microvein in calcite makes microveins complicated and looks like thicker. However, no pure albite growth appeared along the bilateral selvage of microvein located within medium-coarse albite crystal (Fig. 3.74). The albite microvein herein is easily

Fig. 3.72 (+). The width of late stage albite Ab (replacing calcite) may be much greater than that of early albite Ab rim (replacing K-feldspar). Sodium metasomatite in Jiling granite

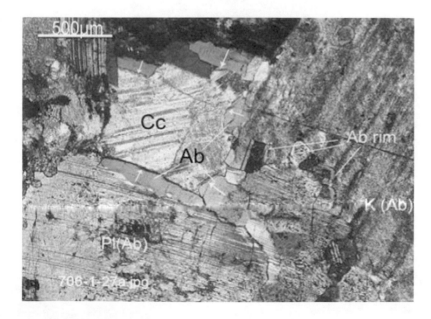

Fig. 3.73 (+)Q. Clear albite rim (shown by *arrows*) grew around the border of albite relicts (*dirty*) with calcite. Sodium metasomatite in Jiling granite

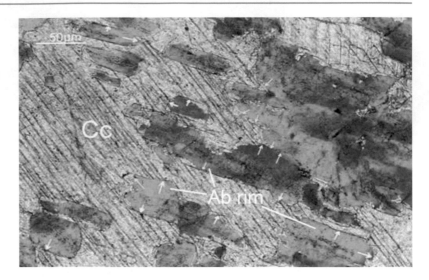

Fig. 3.74 (+). Brown (due to Fe$_2$O$_3$ particles) Ab microveins irregularly cut plagioclase Pl(Ab). No albite growth on the microvein selvage. Sodium metasomatite in Jiling granite

ignored or omitted as it has the same orientation as that of the surrounding medium-coarse albite.

Although the albite microvein and vein selvage albitization are faint and weak, they have a close relation to uranium mineralization, accompanied by the development of vermicular chlorite, according to the research study done by Letian Du and Zhifu Sun (colloquial communication).

Section 3.6.3 shows the rock was in its original unbroken condition while Sect. 3.6.4 implies that it was in the process of being microbroken. So, it is inferred that the late stage clear albite

growth described in Sect. 3.6.3 is probably prior to the albite microvein and selvage albite growth described in Sect. 3.6.4.

3.6.5 Late Stage Quartzification Replacing Calcite

After late stage albitization had ceased, the hydrothermal solution became acidic, producing quartzification that replaces calcite (Fig. 3.77). The late stage quartzification is distributed

Fig. 3.75 (−). Brown
microvein (5-30 μm thick)
with bilateral metasomatic
growth of albite (<100 μm)
in calcite. Sodium
metasomatite in Jiling
granite

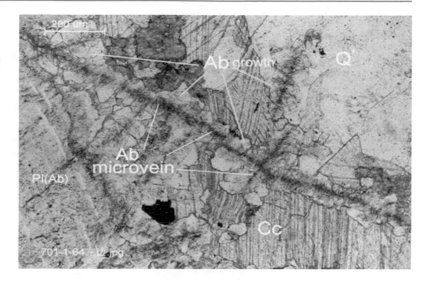

Fig. 3.75 (−). Brown
microvein (5-30 μm thick)
with bilateral metasomatic
growth of albite (<100 μm)
in calcite. Sodium
metasomatite in Jiling
granite

throughout the scope of calcite. The metasomatic
relicts of calcite may be contained in the late
stage quartz (Fig. 3.77).

The central part of late stage quartz grain is
more or less homogeneous, while its periphery
may display inhomogeneous (fan-shaped)
extinction (Figs. 3.78 and 3.79) which is different from the undulatory extinction due to compression. The late stage quartz may sometimes
show euhedral growth line (Fig. 3.80), and may
or may not contain impurities, vermicular or
spherulitic chlorites.

Late stage quartz did not replace albite rim
(Figs. 3.77 and 3.81).

3.6.6   Albite Microvein and Selvage Albite Growth in the Late Stage Quartz Are Residue of Replacement

Figure 3.82 shows that the late stage quartz Q′
was cut by an albite microvein (<10–20 μm

Fig. 3.76 (+)Q. The albite
microvein with the bilateral
metasomatic albite are in
Cc, as well as in late stage
Q′ (*upper right*). (+).
Sodium metasomatite in
Jiling granite

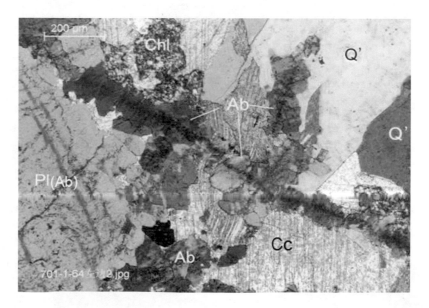

Fig. 3.77 (+). Late stage quartz Q' replaced calcite Cc while the albite clear rim was untouched. Sodium metasomatite in Jiling granite

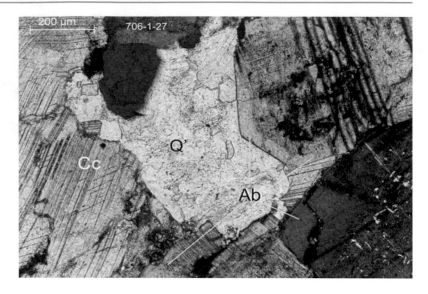

wide) and, seemingly is also replaced by selvage albite growth (up to 50 μm wide).

As far as we know that the most easily replaced minerals are K-feldspar, calcite and tiny perthitic albite, while quartz is the most difficult to be replaced during hetero-oriented albite replacement, except replaced by calcite. Quartz cannot be replaced by albite commonly. However, according to Fig. 3.82, surely it does not mean that albite may also replace quartz?

Careful microscopic observation shows the albite microvein in calcite is always continuous and the bilateral albite growth crystals are more plentiful and well shaped (Figs. 3.75 and 3.76), while the albite microvein in quartz, though generally continuous, may suddenly be stopped somewhere (Fig. 3.83), though locally.

In addition, the contour of the albite crystal growing on microvein selvage in quartz is somewhat smooth. However, it is wondered

Fig. 3.78 (+). Fan-like extinction of late stage quartz Q'. Sodium metasomatite in Jiling granite

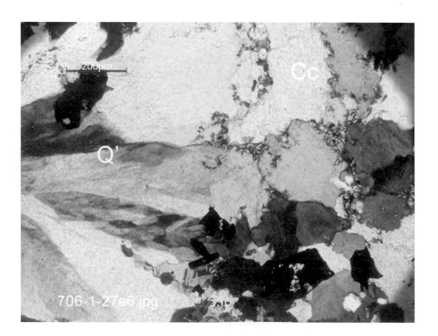

Fig. 3.79 (+). Fan-like extinction of late stage quartz grains Q' with straight and even boundary. Sodium metasomatite in Jiling granite

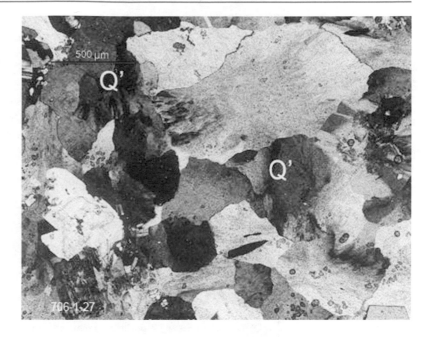

Fig. 3.80 (+). In late stage quartz Q' appears sometimes euhedral growth line implying probably the quartz Q' was formed by filling. Sodium metasomatite in Jiling granite

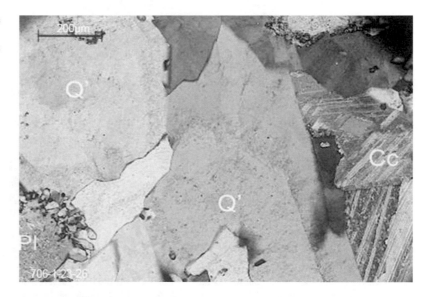

whether albite can replace quartz along the bilateral selvage of microvein, since quartz is hardly replaced by albite according to the authors' observation. If albite microvein was surely formed later than the late stage quartz, the former should maintain continually without unexpected interruption. Of course, microvein does not continue endlessly and may pinch-out naturally. However, normal pinch-out is quite different from sudden interruption. The authors firmly believe that the sudden interruption of the microveins in late stage quartz is abnormal. It should be judged as the result of the replacement of albite by quartz rather than the replacement of quartz by albite on the contrary. It means that during intense late stage quartzification, when calcite is fully replaced and eliminated, small albite grains may also be replaced by quartz as well, resulting in the

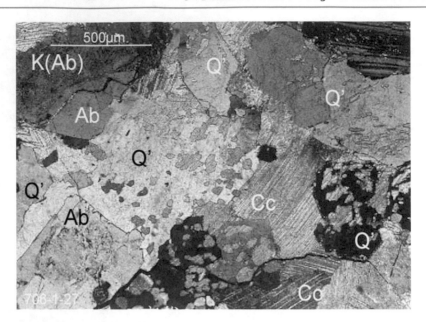

Fig. 3.81 (+). Late stage quartz Q′ replaced calcite and kept albite Ab remained. Sodium metasomatite in Jiling granite

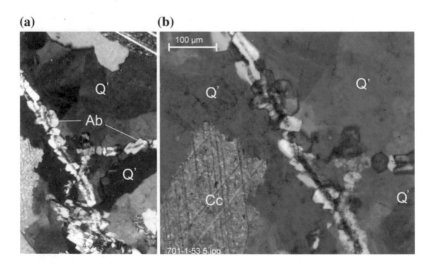

Fig. 3.82 A (+); B(+)Q. Albite microvein and metasomatic growth in late stage quartz. Were the albite microveins formed later than the late stage quartz Q′?

Was the bilateral metasomatic albite formed by replacing Q′? Sodium metasomatite in Jiling granite, Gansu Province

sudden interruption of albite microvein and the corroded (smooth) contour of bilateral albite grains. Therefore, the albite grains are the relicts left over in replacive quartz. So the phenomenon shown in Fig. 3.82 is a false impression.

Therefore, the following six successive sequences of metasomatic processes can be determined in Jiling alkali metasomatite, Gansu Province.

(1) Hetero-oriented myrmekitization and albitization widely spread in granite intrusion body, but weakly affected.

(2) Co-oriented albitization of K-feldspar occurred locally, forming zoned tongue-formed

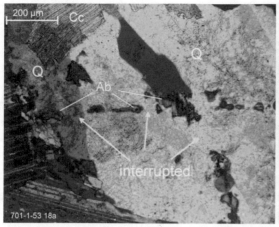

Fig. 3.83 (+)Q. Sudden interruption of albite microvein and corrosion mode of albite grains in late stage quartz, showing that the albite microvein and grains are formed earlier and subjected to the late stage quartzification. Sodium metasomatite in Jiling granite

alkali metasomatites. The albitization is so intense that all plagioclase and nearly all K-feldspar crystals were co-orientedly transformed into albite (Figs. 1.77, 1.78, 1.80 and 1.81) only with a small residual part of K-feldspar left over.

(3) Hetero-oriented calcitization (replacing quartz and K-feldspar) gave rise to the disappearance of whole quartz and most of the K-feldspar (Figs. 1.57 and 1.59) with albite, especially whole albite and chlorite remained. The quartz was even dissolved away and vug (cavity) may be formed, resulting in dissolution-vug-filling textures (see Sect. 3.7.2.4).

(4) Late stage hetero-oriented albitization locally replaced calcite (Figs. 3.69 and 3.72).

(5) Albite microvein (crossing calcite) and bilateral selvage albite growth (replacing calcite) (Figs. 3.75 and 3.76).

(6) Late stage hetero-oriented quartzification (replacing calcite) locally in alkali metasomatite (Figs. 1.50, 3.77 and 3.81) (including the replacement of albite microvein and selvage albite grains).

Figure 3.84 shows that co-orientedly albitized albite contacts quartz and clear albite appears at their border. It is extraordinary that clear albite rim is unexpectedly present at the border of feldspar with quartz. In addition, the feature of the quartz is also abnormal. It should be deduced that the quartz is probably not of primary. According to the above-mentioned multi-replacement processes, it can be inferred that the rock suffered the sequential metasomatic processes as follows:

① K-feldspar was co-orientedly albitized totally to albite;

② Primary quartz was replaced thoroughly by calcite, so calcite directly bordered the co-orientedly albitized albite;

③ Late stage albite replaced calcite, forming clear idiomorphic albite rim;

④ After transformation of hydrothermal fluid from alkaline to acidic, the late stage quartz thoroughly replaced calcite, contacting directly the late stage albite rim.

That is how the texture appeared in Fig. 3.84 was formed.

There is an alternative possibility. Please see dissolution-vug-filling texture in Sect. 3.7.2.4. (2).

Fig. 3.84 (+). K(Ab)—
co-oriented albite K(Ab)
from K-feldspar; *Ab′*—
clear albite rim; *Q′*—late
stage quartz. The rock was
subjected successively to:
the following processes: *1*
K was co-oriented albitized
into K(Ab); *2* Quartz was
totally replaced by Cc; *3* Cc
was replaced by Ab′ rim; *4*
Most Cc was replaced by Q
′. Sodium metasomatite in
Jiling granite

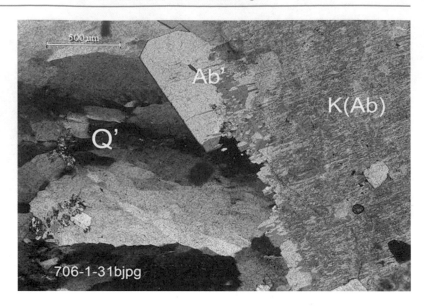

3.7 Distinction Between Metasomatic and Other Textures

Generally speaking, a metasomatic texture can be distinguished from other textures, especially magmatic. However, confusion may be arisen in some places.

The presence of metasomatic relicts of replaced mineral in a replacive mineral may serve as an important evidence for a metasomatic texture. However, several small isolated and co-oriented mineral grains B located in mineral A may not be metasomatic relicts. They may be caused by other reasons. The non-metasomatic textures are also divided into two kinds: co-oriented and hetero-oriented.

3.7.1 Co-oriented Non-metasomatic Texture

Co-oriented non-metasomatic textures are mainly feldspar textures, i.e. intimate intergrowths of K-feldspar with albite, which means that the two kinds of feldspars epitaxially enclosed each other, forming one crystal. The two kinds of

feldspar have the same crystallographic orientation. There are several possible forms.

3.7.1.1 Perthite and Antiperthite
They are the most abundant intimate feldspars intergrowths. Vein-like perthitic albite looks like metasomatic. However, it is formed by simultaneous growth and unmixing of solid solution (see above-mentioned Sect. 3.3 origin of perthite).

3.7.1.2 Rapakivi Texture
Idiomorphic or rounded K-feldspar phenocryst, a few centimeters in diameter, surrounded by a mantle of sodium plagioclase (generally oligoclase) in granitic matrix is called rapakivi texture (Fig. 3.85). Both K-feldspar and oligoclase have the same crystallographic orientation. The mantling plagioclase has the same orientation as that of the perthitic albite and has slight different interference color compared with the perthitic albite.

3.7.1.3 Enclosed or Half-Enclosed Crystal of Plagioclase by K-Feldspar
Similarly, idiomorphic plagioclase may be enclosed or half enclosed by K-feldspar (Figs. 3.86, 3.87). They have similar optical

Fig. 3.85 (+)Q. A corner of Rapakivi texture. The mantling plagioclase Pl (An$_{13}$) has the same orientation of that of the perthitic albite (An < 10). The latter has higher interference color (*yellowish green*). Aqishan granite No. 2, Shanshan County, Xinjiang Autonomic Region

Fig. 3.86 A(+); B(+)Q. Idiomorphic plagioclase Pl is epitaxially mantled by K-feldspar K. Weiya hornblende biotite granodiorite, Hami County, Xinjiang Autonomous Region

(a)   **(b)**

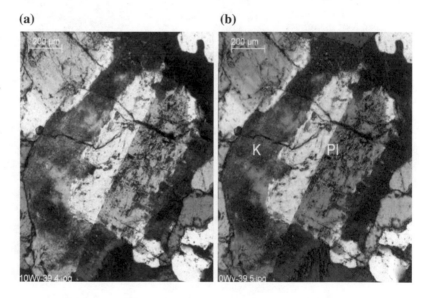

orientation, indicating that they have the same crystallographic orientation.

3.7.1.4 Irregular Contact of Plagioclase with K-Feldspar

K-feldspar may epitaxially crystallize on plagioclase (oligoclase-albite), causing an irregular contact between them (Fig. 3.88). K-feldspar seems to be filled with perthic albite derived from plagioclase. Figure 3.89 shows that a

plagioclase crystal contacts co-orientedly with K-feldspar and a small block (patch) of K-feldspar is enclosed in plagioclase. Does it mean that plagioclase (albite-oligoclase) replaces K-feldspar? Or K-feldspar replaces plagioclase? Since it is a general case or an common phenomenon which is observed everywhere within the whole granite body, the authors believe that is primary rather than metasomatic.

Fig. 3.87 A(+); B(+)Q.
Two euhedral interlocking
crystals of K-feldspar with
albite. K_1, K_2 have the
same orientation as that of
Ab_1, Ab_2. They are primary
rather than metasomatic.
Xihuashan granite, Jiangxi
Province

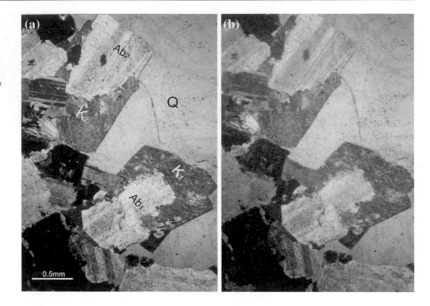

Fig. 3.88 (+)Q. Epitaxial
boundary of Olig-Ab with
K-feldspar, as if the former
co-orientedly replaces
K-feldspar. In fact, it is
similar to rapakivi texture.
No clear rim is formed at
the contact. Naqin
leucogranite

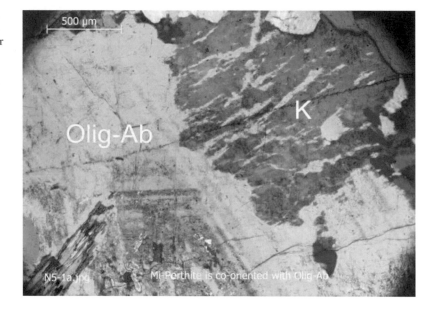

3.7.1.5 Vein-Like K-Feldspar in Co-oriented Plagioclase

The most striking example (Fig. 3.90) shows that
an irregular "vein-like" K-feldspar seemingly
penetrates a co-oriented plagioclase (An_{20-50}). It
is hard to be explained as either being formed as
a normal magmatic inclusion or the result of
unmixing of solid solution. Morphologically, it is
easily treated as an evidence of co-oriented
replacement of plagioclase by K-feldspar. However, the authors still take a skeptical attitude
because this is an individual rare phenomenon
noticed in thin section. Small blocks (patches) of
K-feldspar are also enclosed in the plagioclase
shown in the same Fig. 3.90. There are plagioclase crystals in irregular form that are
co-orientedly enclosed by K-feldspar (Fig. 3.91),
and small blocks of K-feldspar are contained in
many plagioclases in the same thin section.

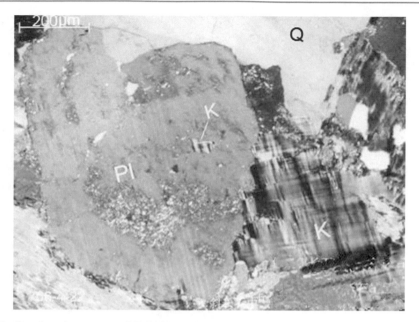

Fig. 3.89 (+). Co-orientation contact of K-feldspar with plagioclase Pl. Pl includes blocks of K-feldspar, as if the former replaces the latter. Jiling granite

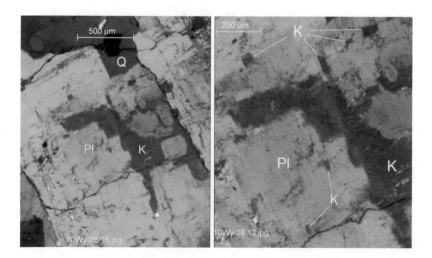

Fig. 3.90 (+)Q. K-feldspar K thrusts into plagioclase Pl ($An_{20—50}$) as if the latter is replaced by the former. The rare phenomenon is still doubted because the Pl also contains the tiny blocks of K. They all have the same orientation. Weiya hornblende biotite granodiorite

Could it be considered that K-feldspar is replacive and at the same time is also a replaced mineral in one plagioclase (Fig. 3.90)?

The authors suggest that oligo-andesine (An_{20-50}) with zoned structure may not be replacive. The phenomenon that K-feldspar "penetrating" irregularly in plagioclase (Fig. 3.90) might be an extreme manifestation in a thin section of interlocking crystals.

Both interlocking crystal of K-feldspar with plagioclase and perthite with perthitic lamellae are easily confused in some places with partly co-oriented albitization of K-feldspar. The former, non-metasomatic co-oriented textures, are

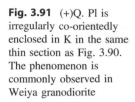

Fig. 3.91 (+)Q. Pl is irregularly co-orientedly enclosed in K in the same thin section as Fig. 3.90. The phenomenon is commonly observed in Weiya granodiorite

generally widely distributed in whole intrusive body, while the latter, metasomatic co-oriented texture is basically characterized by the special shape and distribution occurrence (refer to Figs. 1.68, 1.69, 1.70, 1.71, 1.72, 1.73, 1.74, 1.75, 1.76 and 1.77).

3.7.2 Hetero-Oriented Non-metasomatic Texture

Hetero-oriented non-metasomatic texture means that the texture is formed by intergrowth or close coexistence of minerals of different kinds or orientations due to simultaneous crystallization or another special process. The texture is alike to metasomatic. It can be divided into the following subtypes.

3.7.2.1 Intergrowth or Cotectic Texture

Intergrowth or cotectic texture of alkali-feldspar (mainly K-feldspar, partly also sodic plagioclase) with quartz generally occurs in latest stage of magmatic crystallization mostly in residual magma.

Micrographic texture is easily distinguished from a texture formed by metasomatism by the special pictographic form of quartz (Fig. 3.92) or by granophyric texture in granophyre (Fig. 3.93).

However, the cotectic quartz in granophyric texture may be in rounded or vermicular form (Figs. 3.94, 3.95 and 3.96). In addition, the quartz grains may be separated into several groups, each having its own orientation, which may be misconceived as either K-feldspathization replacing several quartz crystals or quartzification replacing K-feldspar punchingly, namely as "metasomatic punching texture".

Figures 3.97 and 3.98 show the irregular contact between quartz and K-feldspar (perthite). Figure 3.98 even looks like the branch-like K-feldspar replaces quartz. The zigzag boundary might evoke a controversy over whether the texture is formed by metasomatic or by magmatic processes.

The authors have confidence to prove that K-feldspathization is unable to replace quartz since K-feldspatization is always blocked when contacting quartz. So the possibility of replacement of quartz by K-feldspar should be excluded. Quartzification is surely and easily able to

Fig. 3.92 Graphic texture of simultaneous crystallization of K-feldspar and quartz in pegmatite

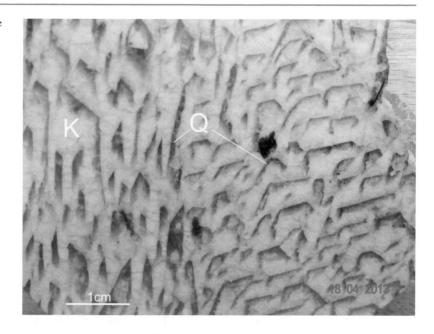

Fig. 3.93 (+)Q. Granophyric texture consists of K-feldspars K_1K_2 growing epitaxially around idiomorphic plagioclases Pl_1 Pl_2, and simultaneously crystallizing quartz Q

replace K-feldspar, but hardly to replace albite, so quartzification would preferentially replace K-feldspar phase with perthtic albite remained. Now it is not the case. The case is that perthite with perthitic albite remained as a whole and especially the fine string of K-feldspar is left over as relic. So it is doubtfully to judge that K-feldspar herein is replaced by quartz.

Therefore, it can be inferred that the phenomenon is still close to a cotectic texture which is probably caused by crystallization at the latest stage in residual magma.

Fig. 3.94 (+)Q. Cotectic or simultaneous crystallization of K-feldspar with quartz. The grain size of quartz became bigger outwardly

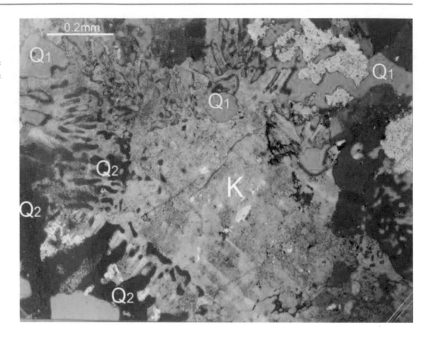

Fig. 3.95 A(+); B(+)Q. Simultaneous crystallization of K-feldspar with quartz. The texture cannot be misconceived as "metasomatic punching texture" or that quartz is partly replaced by K-feldspar

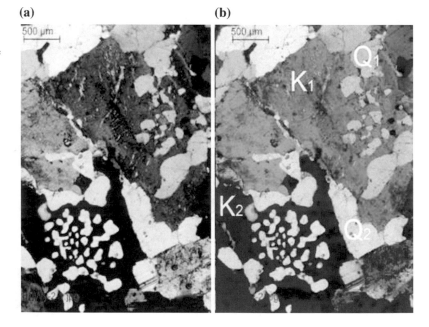

3.7.2.2   Indirect (Inheritable) Replacement Texture

Indirect (inheritable) replacement texture means that a replacive mineral A replaces mineral B after B has replaced mineral C. As a result, it looks as if mineral A directly replaces mineral C.

In fact, mineral A indirectly or inheritably replaces mineral C.

Figure 3.99 shows that quartz is irregularly replaced by both calcite and albite. Quartz is actually replaced by calcite. Is the quartz here also replaced by albite? Is it in contradiction with

Fig. 3.96 (+)Q.
Simultaneous
crystallization of
K-feldspar with three
groups of quartz

Fig. 3.97 A(+); B(+)Q.
Simultaneous
crystallization of quartz
with perthite. Perthitic
albite Ab and K-feldspar K
are interlocked together
rather than separated by
part of quartz

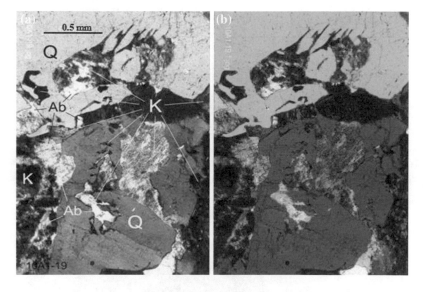

the ordinary rule that albite is unable to replace quartz?

We have to deal with it carefully and seriously. Let us observe extensively the contact situation between plagioclase and quartz in the rock. There is generally no metasomatic phenomenon (i.e., quartz replaced by albite) at the predominant border of plagioclase with quartz even in the same thin section (Fig. 3.100), where calcitization lacks or develops poorly.

The phenomenon that albite looks as if it replaces quartz appears only at the local contact where quartz is irregularly and netlikely replaced by calcite, i.e., quartz relics are enclosed in calcite (Figs. 3.99, 3.101). Since calcite is irregularly distributed at border of plagioclase and albitization is able and easy to replace calcite, it is reasonably referred that "quartz replaced by albite" is not a correct judgement, but a false impression. In fact, the mineral replaced by albite

Fig. 3.98 (+). Cotectic
texture of K-feldspar with
quartz at final stage of
crystallization. Branch-like
K-feldspar penetrates
quartz as if the former
replaces the latter

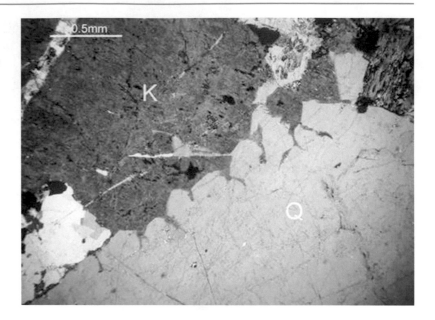

Fig. 3.99 (+)Q. Calcite
replaced quartz and the
unsericitized albite located
at the border of plagioclase
(sericitized) also seemingly
'replaced' quartz.
However, the albite did not
replace quartz, but calcite
instead, in fact. Sodium
metasomatite in Jiling
granite

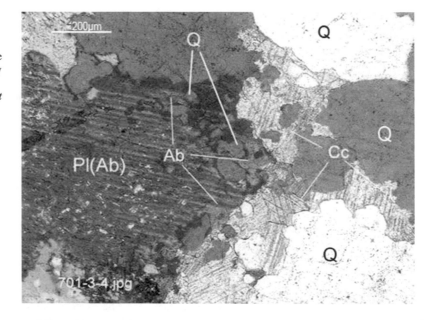

is not quartz, but calcite. In other words, calcite
irregularly replaces quartz at first. Then, at the
contact of plagioclase with netlike calcite, the
newly formed albite may easily replace calcite
(Fig. 3.72), resulting in a false phenomenon, as if
albite has replaced quartz. As a matter of fact, the
albite no more than indirectly or inheritably
replaces quartz (Figs. 3.99 and 3.102).

3.7.2.3 Coarse Myrmekite

Coarse myrmekite means that the vermicular
quartz grains contained in myrmekite are coarse
(Fig. 3.20).

These quartz grains may be divided into several groups and each group has its own orientation. It looks as if the vermicular quartz grains
are relicts from several quartz crystals replaced

Fig. 3.100 (+). The same thin section as Fig. 3.99. No albite rim occurred along the predominant contact of plagioclase with quartz where calcite is absent. Sodium metasomatite in Jiling granite

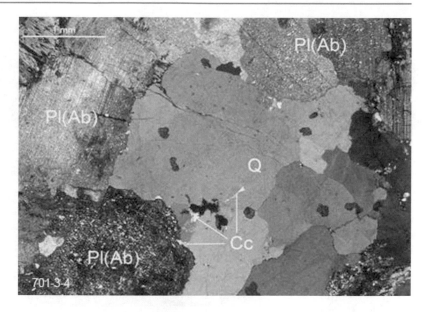

Fig. 3.101 (+). The contact relationship of plagioclase with quartz is evidently complicated only when calcite is present at the border. Sodium metasomatite in Jiling granite

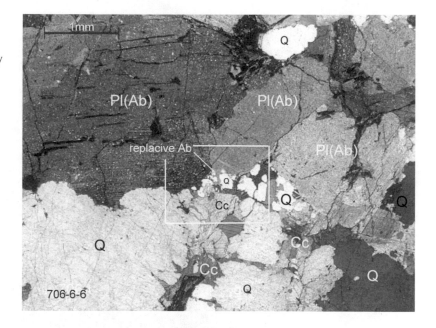

by plagioclase. Actually, the vermicular quartz grains are formed from excess silica after nibble replacement of K-feldspar by plagioclase.

3.7.2.4   Dissolution-Vug-Filling Texture

Dissolution-vug-filling process means that after some parent minerals have been fully dissolved away, an open cavity (vug) appears in rock until precipitation of newly formed minerals occur in it.

Dissolution-vug-filling process is generally noticed in alkali metasomatite, which is called "épisyenite" in French literatures. Episyenitization is characterized by albitization, disappearance of primary quartz, and precipitation of authigenic minerals, such as albite, K-feldspar,

Fig. 3.102 (+).
Enlargement of the frame
in Fig. 3.101. Albite has
seemingly replaced quartz.
In fact, albite has replaced
the nearest calcite, not
quartz

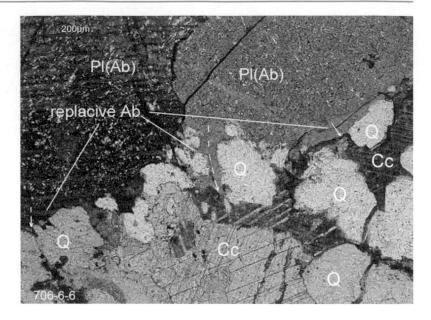

Fig. 3.103 (+).
Medium-grained
porphyritic biotite
monzogranite. Jiling,
Gansu Province

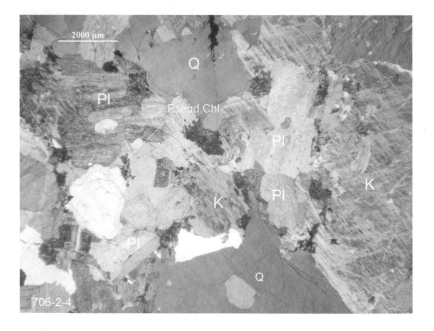

chlorite, illite, calcite, muscovite, quartz, pyrite, anatase, Nb-Ti-Y oxide and other metallic minerals. Special attention has been given to épisyenites in Western Europe, China and North America, where épisyenites act as one of the profitable host rocks for hydrothermal U mineralization.

Let us examine the alkali metasomatite in Jiling medium-coarse porphyritic granite (Fig. 3.103), Gansu Province, North China for example.

Fig. 3.104 (+). Collapse texture after dequartzification. Jiling, Gansu Province

Dissolution-vug-filling process may form: **(1)** Collapse texture and **(2)** Filling texture.

(1) **Collapse texture**

Collapse texture results from accumulation of broken mineral grains (Fig. 3.104). In the empty space, hydrothermal calcite crystals are cemented together (Fig. 3.105).

The collapse space should be greater than the cavity, but the exact location of the cavity possessed by primary quartz cannot be determined.

The broken mineral grains in a collapse place are mainly composed of slightly sericitized plagioclase. These detrital minerals as well as the calcite that partly fill the space are not metasomatic in origin. Only the aggregates of pure albite gathered lumpily at a place occupied probably by primary quartz (Fig. 3.106) are newly grown either in free space after dequartzification or by replacing the calcite which had replaced quartz before.

Fig. 3.105 (+). Calcite fills in vug of collapse texture. Sodium metasomatite in Jiling granite

Fig. 3.106 (+). The space of primary quartz enclosed by dash line is now occupied by pure albite (unsericitized). Sodium metasomatite in Jiling granite

Fig. 3.107 (+). Calcite (Cc) accompanied with shuttle-like hematite (Hm) and oolitic chlorite (Ool.Chl) filled in vug after dequartzification. Later on, clear albite Ab replaced calcite. Sodium metasomatite in Jiling granite

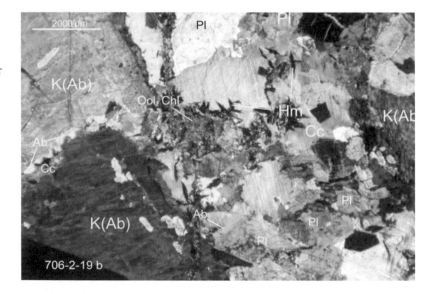

(2) Filling texture

Filling texture occurs where a vug (cavity) is not collapsed after the parent mineral (quartz) has been dissolved away. Later, some minerals precipitate in the vug.

If only one mineral (for example, calcite) fills the vug, it is difficult to distinguish the filling texture from the complete metasomatic texture.

If the vug is filled mainly by calcite, simultaneously accompanied by shuttle-like hematite, aggregates of oolitic chlorite, rare euhedral anatase, apatite and other metallic minerals, it is a dissolution-vug-filling texture (Figs. 3.107, 3.108 and 3.109).

Later on, calcite may easily be replaced by pure albite (Fig. 3.73, 3.74, 3.75 and 3.76) and afterwards, by late stage quartz (Fig. 3.110).

The late stage quartz may totally replace calcite with oolitic chlorite (Figs. 3.111) and albite (3.77, 3.81 and 3.84) remained.

The phenomena shown in Fig. 3.84 might probably include the dissolution-vug-filling

Fig. 3.108 (+). Calcite Cc with oolitic chlorite (Ool. Chl) filled in the vug after dequartzification. Sodium metasomatite in Jiling granite

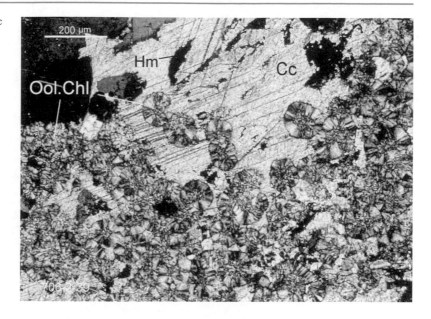

Fig. 3.109 (+). Anatase (Ti) and oolitic chlorite (ool.chl) in calcite (Cc) after leaching of quartz. Sodium metasomatite in Jiling granite

process. It means that primary quartz might be dissolved away and the calcite filled in the vug. Calcite might also be dissolved away and then the late stage quartz with fan-like extinction might fill in the vug. The euhedral contour of the upper part of the clear rim was seemingly crystallized in free space, while its middle and lower part was formed by replacement of calcite.

Cathodoluminescent images in some examples perhaps may show subtle differences between the calcite of filling origin and the calcite of metasomatic origin, as the former displays euhedral growth lines (Fig. 3.112), while the latter does not (Fig. 3.113). However, it is not all the case. It means the cases are not definitely the same.

Fig. 3.110 (+). Calcite Cc is replaced by late stage quartz Q'. Sodium metasomatite in Jiling granite

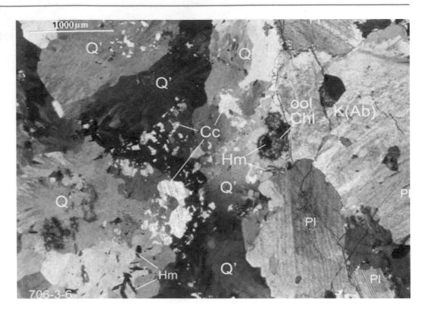

Fig. 3.111 (+). Late stage quartz Q' completely replaced calcite with oolitic chlorite remained

Many researchers believe that quartz leaching is closely related to or immediately follows albitization. Based on the study of fluid inclusion, the investigators considered that the quartz was dissolved by hydrothermal fluids of low-moderate salinity mixed with deep circulated meteoritic water, penetrating into slightly broken granites at low pressure (0.3–1.5 Kb) and at drastic changes of temperatures (450–250 °C) with higher oxygen fugacity (Cathelineau 1986; René 2005; Ahmadipour et al. 2012).

According to microscopic observation, primary quartz may still exist even after the whole feldspar minerals (K-feldspar and plagioclase) have been co-orientedly albitized (Fig. 3.71). So, dequartzification should occur later than

Fig. 3.112 Euhedral
growth line of calcite
(filling type) may be
displayed in
cathodoluminescent image
(*left*). Sodium metasomatite
in Jiling granite

Fig. 3.113 No growth
line is shown in CL image
(*left*) of the calcite which
has replaced Q_1 and Q_2.
Sodium metasomatite in
Jiling granite

complete co-oriented albitization. In addition, the
replacement of quartz by albite has not been
identified so far. Therefore, it is doubtful that
albitization directly leads to dequartzification.
However, quartz and K-feldspar are surely
replaced by calcite (Figs. 3.66, 3.67, 3.68 and
3.69). The calcite precipitated in moderate alkaline solution, in turn, is easily replaced by late
stage albite (Figs. 3.72 and 3.73), and afterwards
is also replaced by newly formed quartz
(Figs. 3.81, 3.110 and 3.111).

Perhaps dissolution-vug-filling process may
be treated as a strong metasomatic process.

The above-mentioned phenomena are quite
easily confused with metasomatic texture. So we
have to observe the contact relationship between
minerals carefully and comprehensively in order
to exclude some non-metasomatic textures.

References

Ahmadipour H, Rostamizadeh G (2012) Geochemical
aspects of Na-Metasomatism in Sargaz granitic intrusion (South of Kerman Province, Iran). J Sci Islamic
Repub Iran 23(1):45–58
Alling HL (1938) Protonic perthites. J Geol 46:142–165

Andersen O (1928) The genesis of some types of feldspar from granitic pegmatites. NGT 10:116–207

Ashworth JR (1972) Myrmekite of exsolution and replacement origins. GM 109:45–62

Aubert G, Burnol L (1964) Observations sur les mineralisations en beryllium du massif granitique d'Echassiers decouverte de herderite. Academie des Sience (Paris) Comptes Rendus 258:273–276

Augustithis SS (1973) Atlas of the textural patterns of granites, gneisses and associated rock types. Elsevier, Amsterdam, p 447

Barker DS (1970) Compositions of granophyre, myrmekite and graphic granite. Geol Soc Am Bullet 81:3339–3350

Becke F (1908) Über Myrmekit. Tschermaks Mineral. Petrogr.Mit 27:377–390

Beus AA, Severov EA, Sitnin AA et al. (1962) Albitized and greisenized granites (apogranites). Nauka, Moscow, pp 1–196 (in Russian)

Burnham CW, Davis NF (1971) The role of H_2O in silicate melt:I.P-V-T relations in the system NaAlSi$_3$O$_8$-H$_2$O to 10 kilobars and 1000 °C. Am J Sci 270:54–79

Burnham CW, Davis NF (1974) The role of H2O in silicate melt: II. Thermodynamic and phase relations to the system NaAlSi$_3$O$_8$-H$_2$O to 10 kilobars and 1000 °C. Am J Sci 274:902–940

Burnol L (1974) Geochimie du beryllium et types de concentration dans les leucogranites du massif central francais. Bureau de Recherches Geologiques et Minieres Memoires 85:137–168

Cahn JW (1968) Spinodal decomposition. Trans Metall Soc AIME 242:166–180

Cathelineau M (1986) The hydrothermal alkali metasomatism effects on granitic rocks: quartz dissolution and related subsolidus changes. J Petrol 27:945–965

Cesare B, Marchesi C, Connolly JAD (2002) Growth of myrmekite coronas by contact metamorphism of granitic mylonites in the aureole of Cima di Vila, Eastern Alps, Italy. J Metamorph Geol 20(1):203–213

Chen Jun J, Lu JJ, Chen W et al (2008) W-Sn-Nb-Ta-bearing granites in the Nanling Range and their relationship to Metallogengesis. Geol J China Univer 14(4):459–473 (in Chinese)

Collins LG (1988) Hydrothermal differentiation and myrmekite—a clue to many geologic puzzles. Theophrastus Publications, Athens, p 382

Collins LG, Collins BJ (2002) K-metasomatism and the origin of Ba- and inclusion-zoned orthoclase megacrysts in the Papoose Flat pluton, Inyo Mountains, California, USA, Myrmekite, ISSN:1526–5757. electronic Internet publication: http://www.csun.edu/~vcgeo005/Nr44Papoose.pdf

Cuney M, Autran A, Bornal L (1985) Premieres resultats apportes par le sondage GPF de 900 m realise sur le granite sodolithique et fluore a mineralization disseminee de Beauvoir. Chron Rech Min 481:59–63

Deer WA, Howie RA, Zussman J (1963) Rock-forming minerals 3.4. Longmans, London

Deer WA, Howie RA, Zussman J (2001) Rock-forming minerals: Framework silicates. Feldspars. vol. 4a. Longmans, London

Dickson FW (1966) Porphyroblasts of barium-zoned K-feldspar and quartz, Papoose Flat, Inyo Mountains, California, Geology and Ore Deposits of the American Cordilleran: Geological Society of Nevada Symposium Proceedings, Reno/Sparks, Nevada, April 1995, p 909–924

Dong SB and He GP (1987) The classification of metasomatic replacement fabrics of granitic rocks and its genetics significance. Bull Chin Acad Geol Sci 16:71–79 (in Chinese)

Drescher-Kaden FK (1948) Die Feldspat-Quartz-Reactionsgefüge der Granite und Gneise und ihre genetische Bedeutung. Springer, Heidelberg, p 259

Du S, Huang Y (1984) On study of Xianghualingite. Sci Sinica, part B (11):1039–1047

Eadington PT, Nashar B (1978) Evidence for the magmatic origin of quartz-topaz rocks from the New England Batholith. Aust Contrib Min Petro 67:433–438

Expedition Nanling Regional Geological (1959) Preliminary reaearch on nanling intrusions. Geological Publishing House, Beijing (in Chinese)

Fenn PM (1977) The nucleation and growth of aklaki feldspars from hydrous melts. Can Mineral 15:135–161

Harker A (1950) Metamorphism. Methuen, London

Hong W (1975) Characteristics and metallogenesis of Ta, Nb REE-bearing granites in China. In: National rare element conference proceedings (first set). Science Press, Beijing, pp 50–62 (in Chinese)

Hu S (1975) Igneous rock of sodic and potassic series and alkali metasomatism to metallogenesis for rare elements. In: National rare element conference proceedings (first set). Science Press, Beijing, pp 91–94 (in Chinese)

Hu SX (1980) Petrography of Metasomatically Altered rocks (Research Guide of rock slices). Geological Publishing House, Beijing (in Chinese)

Hubbard FH et al (1965) Antiperthite and mantled feldspar textures in charnokite (enderbite) from S W Nigeria. Am Mineral 50:2040–2051

Imeopkaria EG (1980) Ore-bearing potential of granitic rocks from the Jos-Bukuru complex, Northern Nigeria. Chem Geo 28:67–70

Kovalenko BE, Kuzmin ME, Antapin BS (1971) Topaz-bearing quartzic keratophyre (ongonite) of subvolcanic vein magmatic rocks. Rep Acad Sci USSR 199(2):119–121

Lehmann J (1885) Uber die Mikroklin- und Perthitstruktur der Kalifeldspathe und deren Abhangigkeit von ausseren zum Theil mechanischen Einflussen. Jahresber. Schles. Ges. Vateri. Cultur 63:92–100, 64:119 (1886)

Liu Y, Li H, Lin D et al. (1975) Space distribution characteristics of endogenetic rare element deposits in China. In: National rare element conference proceedings(first set). Science Press, Beijing, pp 1–23 (in Chinese)

Lofgren GE, Gooley RG (1977) Simultaneous crystallization of feldspar intergrowths from the melt. Am Mineral 62:217–228

Luth WC (1976) Granitic rocks. In: Bailey DK, Macdonald R (eds) The evolution of the crystalline rocks. Academic Press Inc., New York, p 484

Masgutov RV (1960) Yisvestia Akademi Nayk. Geol Ser Tom 3 (in Russian)

Michel-Lévy A (1874) Structure microscopique des roches acides anciennes. Granite poprphyroide de Vire. Bull Soc géol France 3:201

Parson I, Lee RE (2009) Mutual replacement reactions in alkali feldspars I: microtextures and mechanisms. Contrib Mineral Petrol 157:641–661

Phillips ER (1974) Myrmekite-one hundred years later. Lithos 7:181–194

Phillips ER, Ransom DM (1968) The proportionality of quartz in myrmekite. Am Mineral 53:1411–1433

Plümper O, Putnis A (2009) The complex hydrothermal history of granitic rocks: multiple feldspar replacement reactions under subsolidus conditions. J Petrol 50:967–987

Pryer LL, Robin PYF (1996) Differential stress control on the growth and orientation of flame perthite: a palaeostress-direction indicator. J Struct Geol 18:1151–1166

Raimbault L (1984) Geologie, petrographie et geochimie des granites et mineralization associees de la region de Meymac (Haut Correze, France), Ph.D. Thesis, Paris. Ecole de Mines, p 482

René M (2005) Geochemical constraints of hydrothermal alterations of two-mica granites of the Moldanubian Batholith at the Okrouhlá Radouň uranium deposit. Acta Geodyn Geomater 2(4):63–79

Rong JS (1982) Microscopic observation on the metasomatic phenomena of rock-forming minerals in granites. Petrol Res 1(1) (in Chinese)

Rong JS (1992) Origin of Myrmekite. Acta Petrologica et Mineralogica 11(4):324–331 (in Chinese)

Rong JS (2002) Myrmekite formedbyNa-andCa-metasomatism of K-feldspar. Myrmekite ISSN1526–5757. electronic Internet publication: http://www.csun.edu/ ~ vcgeo005/Nr45Rong1.pdf

Rong J (2009) Two patterns of monomineral replacement in granites. Myrmekite ISSN 1526–5757. electronic Internet publication: http://www.csun.edu/ ~ vcgeo005/Nr55Rong3.pdf

Rosenqvist IT (1950) Some investigations in the crystal chemistry of silicates (II). NGT 28:192–198

Schwantke A. 1909. Die Beimischung von Ca im Kalifeldspat und die Myrmekibildung. Contrib Min Geo 311–316

Sederholm JJ (1897) Uber eine archaische Sedimentformation im sudwestichen Finland. BCGF (6):254

Smith JV (1961) Explanation of strain and orientation effects in perthites. Am Mineral 46:1489–1493

Smith JV (1974) Feldspar minerals(2). Springer Speer, New York, p 1984

Smith JV, Stenstrom RC (1965) Electon-excitedluminescence as apetrologic tool. J Geol 73:627–635

Spencer E (1945) Myrmekite in graphic granite and in vein perthite. MM 27:79–98

Stemprok M (1979) Mineralized granites and their origin, Episodes Geol. News Lett 3:20–24

Swanson SE (1977) Relation of nucleation and crystal growth rate to the development of granites textures. Am Mineral 62:966–978

Taylor RP (1992) Petrological and geochemical characteristics of the peasant ridge zinnwaldite-topaz granite, southern New Brunswick, and comparisons with other topaz-bearing felsic rock. Can Mineral 30:895–921

Tschermark G (1864) Chemisch-mineralogische Studien. I. Die Feldspathgruppe. SAWW 50:566–613

Vernon RH (1986) K-feldspar megacryst in granites— Phenocrysts not porphyroblast. Earth Sci Rev 23:1–63

Vernon RH (1999) Flame perthite in metapelites gneisses at Cooma, SE Australia. Am Mineral 84:1760–1765

Vernon RH, Paterson SR (2002) Igneous origin of K-feldspar megacrysts in deformed granite of the Papoose Flat Pluton, California, USA. Electron Geosci 7:31–39

Vogel TA (1970) The origin of some antiperthites—a model based on nucleation. AM 55:1390–1395

Vogt JHL (1905) XXVI. Physikalisch-chemische Gesetze der Krystallizationsfolge in Eruptivgesteinen. TMPM 24:437–542

Vogt JHL (1926) The physical chemistry of magmatic differentiation of igneous rocks.II. On the feldspar diagram Or:Ab:An. Norsk.Vidensk.-Akad. Oslo I. Mat.-nat. Kl. Skr.(4)

Wang D (1975) Classification and genesis of the rare element mineralized granite in China. In: National rare element conference proceedings (first set). Science Press, Beijing, pp 63–66 (in Chinese)

Wang LK, Wang HF, Huang ZL (1998) The three end members of Li-F granites and their origin of liquid segregation. Chin J Geochem 17(1)

Wang LK, Huang ZL (2000) Liquid separation and experiment of Li-F granite. Science Press, Beijing (in Chinese)

Wang Z, Yu X, Zhao Z et al. (1989) Rare earth element chemistry. Science Press, Beijing, pp 225–245 (in Chinese)

Wang D, Zhou X, Xu X (2002) The K-feldspar Megacrysts in Granites: A case study of microcline megacrysts in Fugang granitic complex, South China. Geol J China Univer vol 2 (in Chinese)

Willaime C, Brown WL (1974) A coherent elastic model for the determination of the orientation of exsolution boundaries: application to the feldspars. Acta Crystallogr A 30:316–331

Willaime C, Brown WL (1985) Orientation of phase and domain boundaries in crystalline solids: discussion. Am Mineral 70:124–129

Xia HY, Liang SY (1991) Genesis series of W, Sn rare metal granite deposit in Southern China. Science Press, Beijing (in Chinese)

Xia W, Zhang J, Feng Z et al. (1989) Geology of Rare Metal Ore Deposits in Granites in Nanling. China University of Geosciences press, Beijing, pp 14–115 (in Chinese)

Yuan Z, Bai G, Yang Y (1987) Discussion on Petrogenesis of rare metal granites. Mineral Deposits 6(1):88–94 (in Chinese)

Zhang J, Xia W (1985) Preliminary study on the geological and metallogenic mechanism of W, Sn, Nb, Ta in Songshugang Deposit. In: Geology and mineral resources of Nanling (first set). Geological Publishing House, Beijing, pp 145–148 (in Chinese)

Zhu J, Liu W, Zhou F (1992) Ongonite in Xianghualing. In: The annual report of State Key Laboratory of metal deposits, Nanjing University. Nanjing University press, Nanjing, pp 12–19 (in Chinese)

Zhu JC, Rao B, Xiong XL et al. (2002) Comparison and genetic interpretation of Li-F rich, rare-metal bearing granitic rocks. Geochimica 31(2):141–152 (in Chinese)

Conclusion

4

After formation of granite, under a new physicochemical environment, some gasses and solutions either relevant or irrelevant to magmatic process can penetrate into solidified granitic rocks, resulting in metasomatism, i.e., the occurrence of partial dissolution of unstable preexisting minerals and local crystallization of more stable secondary minerals, and change of local texture.

Metasomatic textures and phenomena discussed in this book are formed in granitic rock after solidification of magmatic crystallization. Since then, the rock has not been subjected to recrystallization.

On the basis of consistency or difference of crystallographic orientations between replacive and replaced minerals, the authors discriminate replacement textures into two patterns: hetero-oriented replacement and co-oriented replacement. Mostly they are of hetero-oriented replacement. The co-oriented replacement occurs only between minerals with the same or similar crystallographic lattice. In other words, the replacive mineral keeps the same or similar crystallographic lattice as that of the replaced one.

Hetero-oriented replacement occurs at the grain boundary between two minerals with different orientations. Two indispensable background conditions are required: (1) presence of a mineral prone to be replaced on one side; (2) presence of a mineral acting as nucleation center for the replacive mineral on the other side, i.e., the same or similar mineral against which the replacive mineral may lean and epitaxially grow.

Without one of the above two conditions, no hetero-oriented replacement can occur basically.

Impurity may serve as a nucleus center, when metasomatism should occur but there is no same or similar mineral as the replacive mineral aside in the rock, such as calcitization. However, in case of presence of the same or similar mineral as replacive one, the replacive mineral would certainly take the former as a nucleus center rather than a foreign matter or impurity.

Hetero-orientated replacement would not occur at the grain boundary between two minerals of the same or similar kind with the same orientation.

During hetero-oriented replacement the following metasomatic products may be present, including clear albite rim, intergranular albite, myrmekite, newly formed K-feldspar replacing old K-feldspar or plagioclase, newly formed quartz, muscovite, beryl, etc. while the major replaced mineral is K-feldspar. For calcitization, the first replaced mineral is quartz, followed by K-feldspar. Dissolution–precipitation is the mechanism for hetero-oriented replacement.

Co-oriented replacement means that both replacive and replaced minerals have the same or similar orientation. For sheet alumosilicate minerals, such as the co-oriented replacement of biotite by muscovite or chlorite, the ion exchange may probably be the replacement mechanism. For framework alumosilicate (as well as phosphate) minerals, the mechanism of co-oriented replacement has not been identified so far. However, the appearance and enhancement of

J. Rong and F. Wang, *Metasomatic Textures in Granites*, Springer Mineralogy,
DOI 10.1007/978-981-10-0666-1_4

micropores induced by an external hydrothermal fluid may be an important function causing the co-oriented replacement. Perculating into micropores the hydrothermal fluids might dissolve the side unstable mineral and at the same time precipitate stable mineral. As the replacement is proceeded within the mineral of the same or similar kind, the newly formed replacive mineral may take the same orientation as that of the old one. Possibly it fits for co-oriented albitization of plagioclase. As for co-oriented albitization of K-feldspar, the mechanism is not clear yet and needs further investigation.

Inserting a quartz plate under crossed polars accompanied with slow rotation of objective stage is a helpful procedure to determine the consistence of crystallographic orientations of various grains of minerals, especially those with low birefringence, so as to identify either hetero- or co-oriented replacement phenomena.

The small platy albite minerals in Li-F granites have a primary origin rather than metasomatic in "chaotic" type.

Myrmekite is produced from K-feldspar replaced by plagioclase as the SiO_2 content needed to compose replacive plagioclase is less than that contained in the replaced K-feldspar and the surplus SiO_2 remaining in situ produces the vermicular quartz grains.

Perthitic albite lamellae look like metasomatic in origin from a morphological point of view. However, they may be formed by either exsolution (for finer, denser and regular forms) or by simultaneous growth (for coarser and more irregular forms). Flame perthitic albite is likely formed in relation to injection and replacement. The development of chessboard albite formed by co-oriented albitization of K-feldspar starts from group-dotted to slice (along (010)), then assembles to block, lumpy form at a rapid rate.

K-feldspar may be subjected to both hetero-oriented and co-oriented albitization. The former occurred much earlier than the latter. When the former develops intensely, the latter does not happen at all. On the contrary, when the latter strongly develops, the former is still not promoted or enhanced. It implies that they are different metasomatic processes and their formation conditions and environments must be quite different, although the real difference is not clear yet. It is reasonable and necessary to divide them into two kinds and describe them separately rather than lump them together. Their individual formation conditions are worth investigating.

The key evidence of replacement is the presence of relics of the replaced mineral in the newly formed replacive mineral. However, such relics should be distinguished from crystals formed by simultaneous crystallization and from normal inclusions in igneous rocks. The relics also should be determined as real residue from the present replacement rather than an inherited relict from the previous replacement.

Granitic rock may be subjected to multistage metasomatism. According to the metasomatic rule and the contact relationship, it is possible to interpret the history of superimposed metasomatic processes.

Dissolution-vug-filling process does not belong to metasomatic processes in a strict sense. But it is closely accompanied by intense replacement process.

Because of the limitation of thin sections, a false impression may likely occur when an individual phenomenon is noticed and firmly established. Only when a special phenomenon is repeatedly observed in arbitrary thin sections, can we confidently believe it and accordingly make a reliable judgement.

Printed in the United States
By Bookmasters